PERMACULTURE GARDENING MADE EASY

A 7 STEP BEGINNER'S GUIDE TO COMPANION
PLANTING, ORGANIC FARMING, AND BUILDING A
FOOD FOREST IN YOUR BACKYARD

PERENNIAL PUBLISHING

CONTENTS

INTRODUCTION

A lot of people look at farming as a very boring career line, this generation especially. They often regard it as "dirty work," which is usually at the bottom of their shelf when choosing a career.

But is farming really so uninteresting, or do they just have the wrong idea about it? Well, there is one other thing this generation is known for: having a knack for rebellion.

So let's do something really quick. Imagine a farming system that consistently goes against the grain of modern farming. A piece of land where you can cultivate diverse mixtures of plants and animals in place of monocultures. Where there's no need for chemicals because the dynamic interactions between the plants and animals in the polycultures provide all the services that conventional farmers find in the fertilizer bag and the crop-sprayer, THAT is Permaculture farming!

Sounds fascinating, doesn't it? Since you choose to read this book, I should assume so. You've come to this book because you want to learn more—possibly everything—about permaculture farming.

I used to do things the old-fashioned way when I started out. I'll be the first to admit that it was never fun, never exciting. It was all just one big cycle; prepare the grounds, plant the seeds, and nurture. Doing things this way may be standard and productive, but it's only a matter of time before you find yourself asking, "Is this what I'm supposed to do with my life?"

The regular techniques and methods are all but dynamic, and they would quickly have you feeling this way. It was a huge game-changer when a buddy of mine let me on about the little secret that is permaculture farming. I'm glad I had a Master like that to show me the light side; allow me to do the same with you with this book!

Permaculture farming, as you may have guessed, is nothing new. It was created by a group of brave men and women as a response to food insecurity and the desire for independence. Permaculture is a design philosophy that emerged during very dark times; the oil crisis of the 1970s. It involves everything from recycling, reusing, and regenerating to simply observing. It combines mindset and practical application.

In terms of gardening, it implies that not only can we produce food nearly everywhere, from fruit shrubs in patio pots to vines on fences, but that we can also receive larger yields with less work by just copying nature.

The moment I began to see every nook as a potential area for food production, the world became one giant gingerbread house in my eyes.

To be a rebel farmer, you need more than just a contrarian nature. It also requires expertise and knowledge, neither of which are easy to come by. In this book, I share the expertise that I, along with a number of other specialists, have accumulated over the course of our lifetimes. It covers every facet of this farming, including how to establish a holistic system on the farm as well as how to make a livelihood from it.

Everything needed to start your own thriving permaculture farm or garden—from the broad ideas that served as my roadmap to the specifics, including the fruit kinds I have discovered work best for permaculture growing—is included in this book.

What's a rebellion without a rebel base? Learning about your own particular place is a crucial aspect of permaculture. Every region of the Earth has its own personality and character, just like every individual. Here, you will learn how to appreciate and maximize your own farmland.

The book's attitude lessons are more important than the knowledge it contains. Its message is more along the lines of "this is how you thought it was done" than "this is how you do it;" After all, you're living by your own rules now, and you should do things in your own style!

With all the freedom to practice farming your own way, you still have to be able to practically apply your techniques and convert them to productivity. Experience with practical applications is essential in permaculture design. Understanding nature just through theory is challenging. You only truly become a permaculture designer if you have accumulated a lot of hands-on experience over time.

You can also share your newfound knowledge and freedom with others like I am with this book.

In the pages of this book, you will learn actual working, practical experience on how to set up your own permaculture farming system. The step-by-step guide will help you transform your curiosity into farm or garden productivity. Knowledge has been passed down from permaculture experts and rebel veterans; it is yours for the taking and practice. The modern farming rebellion has only just begun, my friend!

1

PERMANENT + AGRICULTURE = PERMACULTURE

"We had a 22-member CSA food box program this year, and I would consider our farm to be practicing permaculture. Half the boxes were for singles and couples, and the other half were for families. I'd estimate that would be about 60 people.

We provided an average of 7 items to our customers for 18 weeks on about an acre of cultivated land and wild edibles.

We also sold extra produce to restaurants and had a weekly table at a nearby farmer's market for about 14 weeks. And of course, we ate a lot of veggies from the garden ourselves.

We probably could've been able to support more than the 22 members but wanted to start small as it was our first year on this farm."

- Travis Phillip

Just one of the many success stories from the light side of the farm.

Still not convinced? No worries, we're just getting started.

Permaculture At A Glance

What really is this permaculture farming that I've been going on about?

In this section, I'll do my best to explain it all to you and open your eyes to the production possibilities.

First off, let's start with a proper definition. The term was originally coined by the Tasmanian son of a fisherman in 1978. Bill Mollison defined "permaculture" as "The conscious design and maintenance of agriculturally productive systems which have the diversity, stability, and resilience of natural ecosystems. It is the harmonious integration of the landscape with people providing their food, energy, shelter, and other material and non-material needs in a sustainable way."

In other words, permaculture is both a scientific method for how to live in harmony with nature and a comprehensive worldview that emphasizes doing so.

When you practice permaculture, you're applying innovative design techniques based on whole-systems thinking and also taking into account all flowing materials and energy that have an impact on or are influenced by suggested alterations.

Let me break it down for you.

It means that both upstream and downstream consequences must be properly taken into account in both the short- and long-term before making changes to overland water flow, for example. Or, while examining an "issue," such as brushy vegetation, one takes into account how its removal or alteration may affect the soil and animals and how these interrelated forces will evolve through time and place.

History of Permaculture

1929s Tree Crops: A Permanent Agriculture is one book I find very interesting, and in a way, it started this all. If you want to know how it all started, then that book has all the answers. The author, Joseph Russell Smith, summarized his extensive experience dealing with fruits and nuts as crops for human sustenance and animal feed. Smith used an antecedent word as the subtitle. He proposed a mixed system of crops and trees underground because he considered the world as one interconnected entity.

This book inspired many people working to improve agriculture's sustainability, including Toyohiko Kagawa, who in the 1930s established forest farming in Japan.

Let's fast forward to the 60s. *Water for Every Farm*, a book by Australian, P. A. Yeomans, endorsed the idea of permanent agriculture as that which may be perpetuated indefinitely.

Yeomans presented an observation-based approach to land use in the 1940s. He also established the keyline design as a method of regulating the supply and distribution of water in Australia in the 1950s.

There were a lot of other big names and influencers throughout the 20th century. Holmgren identified Stewart Brand's early works as an influence. Esther Deans invented no-dig gardening, and Ruth Stout was an additional early influencer. Masanobu Fukuoka wrote *The One-Straw Revolution* in 1975 and began promoting no-till orchards, gardens, and natural farming in Japan in the late 1930s.

All of those accomplishments contributed to the very first practitioners of permaculture. See, it all started in the late 1960s, on the island state of Tasmania in southern Australia's federation. Bill Mollison, along with his partner, David Holmgren, began formulating concepts for sustainable agriculture systems. This was due to the risk posed by the employment of industrial-agricultural techniques, which is constantly expanding.

These brave men believed that these techniques were heavily reliant on non-renewable resources, as well as polluting the air, water, and land, decreasing biodiversity, and eliminating billions of tons of topsoil from formerly fruitful areas of the world. They responded by developing a design methodology known as permaculture, which was first made public in 1978 with the release of their book *Permaculture One.*

Early in the 1980s, the idea would evolve from designing agricultural systems to sustainable human environments. From there, hundreds of permaculture sites were designed after *Permaculture One*.

Mollison went on to write other in-depth volumes, including *Permaculture: A Designers' Manual*, in which he further extended and improved the concepts. Hundreds of students took Mollison's two-week Permaculture Design Course (PDC), which he taught in more than 80 different nations.

Graduates were encouraged by Mollison to become teachers themselves and establish their own institutes and demonstration places. The rapid growth of permaculture depends on this multiplier effect. It was a true turning point for the agricultural industry.

PRINCIPLES OF PERMACULTURE

Permaculture comes with a few principles that enable us to fully appreciate and apply its unconventional methods.

Closed Loop Systems

Sustainable systems, by definition, meet their own energy requirements. This idea can be applied to what permaculturists refer to as "inputs," such as food and fertilizer, in addition to things like biofuels and solar electricity. For instance, the system may be created to meet the farm's or garden's own fertility requirements, perhaps from livestock manure or cover crops, rather than importing fertilizer. Additionally, if you are raising livestock, you should aim to produce all of the food for your animals on-site, whether that means growing grains, forage crops, or animal feed from recycled kitchen scraps. Remember that a successfully closed loop system "turns waste into resources" and "issues into solutions," as any permaculturist worth their salt would say. If you've ever observed how happily ducks consume snails, Mollison's catchphrase, "You don't have a snail problem, you have a duck deficit," would make perfect sense.

Perennial Plants

Not only permies understand that tilling the land even just once or twice a year isn't very beneficial for the soil. Those practicing agroforestry are on the right track. Think of shade-grown coffee or cocoa farms in South America when considering agroforestry. These emphasize the development of edible tree crops along with related understory vegetation. In order to avoid constant tillage, they advise choosing perennial crops, which are ones that are only planted once. There is little doubt that agriculture would be far more sustainable if we could replace all of the monocultures of corn, soy, and wheat with agroforestry systems. The only issue is that very few of the crops that the majority of us eat are perennials.

Multiple Functions

Every element of a structure or landscape should have more than one purpose, according to one of permaculture's more novel concepts. The goal is to build an integrated, self-sufficient system through the thoughtful design and positioning of its parts. For instance, if you need a fence to keep animals in, you may design it to serve as a windbreak, a trellis, and a reflecting surface to increase the amount of heat and light that reaches neighboring plants. In addition to providing water for irrigation, a rain barrel can be used to grow edible fish and aquatic food plants.

Eco-Earthworks

Water conservation is a key priority in permaculture farms and gardens, where the land is frequently skillfully shaped to funnel every last drop of rain toward some useful goal. Terraces on steep terrain, swales (wide, shallow ditches designed to catch runoff and cause it to soak into the ground around plantings), or a network of canals and planting berms on low, swampy territory may all be used to achieve this. The latter is based on chinampas, a method of raising food, fish, and other crops in an integrated system used by the ancient Aztecs. Permaculturists frequently hail chinampas as the most successful and long-lasting type of agriculture ever created.

Let Nature Do the Work for You

The Mollisonian mantras of "working with, rather than against, nature" and of "protracted and attentive observation, rather than protracted and mindless labor" are possibly the greatest representations of the permaculture ethos. On a practical level, these concepts are implemented through the use of items like chicken tractors, where the natural scratching and bug-hunting activity of hens are harnessed to clear an area of pests and weeds in preparation for planting - or simply by planting mashua under your locust trees. While locust trees are renowned for supplying nitrogen to the soil, mashua, an Andean root crop that can withstand shade, need a support system to grow on. Therefore, the locust's inherent qualities prevent the need to deal with fertilizer or construct a trellis to offer shade. Not only that, but it also provides nectar for bees and is aesthetically pleasing. This way, you can maximize hammock time by letting nature take care of the chores of farming and gardening. This is another one of Mollison's well-known maxims, and it sounds much better than constantly laboring to do all of this work yourself, doesn't it?

ETHICS OF PERMACULTURE

Permaculture design is built on the principles of earth care, people care, and fair share, which are also prevalent in the majority of traditional communities. We can better comprehend good and negative outcomes when we see ethics as culturally evolved processes that control self-interest.

Ethics are increasingly important for long-term cultural and biological survival the more powerful humans grow.

Research into community ethics and learning from cultures that have coexisted with their environment in a largely sustainable manner for much longer than more recent civilizations are the sources from which permaculture ethics are distilled.

This does not imply that we should disregard the great modern teachings, but rather that as we move toward a sustainable future, we must take into account beliefs and principles that are different from the conventional norm.

Earth Care

The Earth is the single source of the things we need to survive—air, water, food, and shelter—and it is the very thing that keeps us alive. We can't receive these things from anywhere else. We are entirely dependent on the Earth and all of the planet's living organisms—all of which are related to one another in a complex web of interdependence—for our survival.

It makes sense that protecting the ecosystems on Earth that keep us alive would be considered "enlightened self-interest," or acting morally to secure one's own life. We can achieve this by not polluting the air we breathe and nuturing the land that sustains us.

All living and non-living creatures, including plants, animals, and the soil, water, and air, are considered to be a part of "caring for the earth." Why? Ecology and biology demonstrate this interdependence and connection between all living and non-living systems. All people are impacted when one is.

The soil needs to be taken care of as well as the Earth. Every aspect of life is interdependent, and the soil is essentially a very intricate living ecosystem that sustains plant life. In turn, plant life sustains higher organisms and offers us, directly or indirectly, our sources of food.

Beyond guaranteeing a supply of clean air, taking care of the Earth also entails taking care of our forests, which are the planet's lungs. As a result, forests play a crucial role in securing our supply of fresh water. Forests are also intimately connected to the process of rain creation and the water cycle. It entails taking care of the rivers that run through our world like the veins that carry the water on which all life depends.

People Care

All living things, including humans, are interdependent on one another. In actuality, people are naturally communal and social, as the proverb "no man is an island" states. On this planet, most life is cooperative by its very nature.

If you have any doubts about the truth of this assertion, go back in time, past the industrialized civilization we currently live in, and examine history. Banishment or exile—being driven away from the group and left to fend for oneself—was traditionally the punishment for significant wrongdoing in ancient societies. This was the equivalent of being given the death penalty, or at the very least, a harsh, dangerous, and lonely life. Humans are interdependent not only in a physical sense but also psychologically. Recent research has demonstrated the positive effects of community on mental health, as well as the negative effects of community absence. The nature of punishment reflects the ancients' understanding of the need for fellowship among humans. Unfortunately, modern society has forgotten this, and people have banished themselves to the useless and isolating electronic prison that is modern life, where they don't get out and enjoy life with friends or even meet their neighbors.

Self-sufficiency is a dangerous myth.

"Care of People" aims to encourage interdependence and duty to the larger community. It's important to note that this discussion is about self-reliance, not self-sufficiency. As I've said before, "no man is an island," so it is absurd to expect any one individual to live anything other than the most rudimentary existence to be able to do everything. Bill Mollison once said, "I might grow food, but I don't want to have to make my own shoes; I can trade food I've grown with someone who makes shoes." That is what makes a community! It's about helping one another out and sharing.

So what does it mean to encourage self-reliance? It entails accepting responsibility for more than just one's own future and aiming to support one's community by imparting knowledge and experience to enable others to develop the skills necessary to meet some of their most basic requirements. "Give a man a fish, and he'll eat for a day. Teach a man to fish, and he'll eat forever" sums up this situation in its essence. It's about working together to alter both one's own life and the lives of others.

A stable, caring, and emotionally healthy community is built when people work together to assist one another and meet their needs, both physical and non-physical.

Importantly, "Care of People" must start with the person nearest to us—with ourselves! It's challenging to take care of others when we can't take care of ourselves, and taking care of others while ignoring oneself is pointless. Such martyrdom is unproductive because, if we are interested in assisting others, it is in our best interests to be in the best possible condition to be of service to them. Beyond our own particular self, "Care for People" then encompasses the next group of people in our lives: our families, followed by our neighbors, our immediate neighborhood, the larger community, and finally, all of humanity.

Fair Share

The "Return of Surplus to Earth and People" ethical principle is another name for this.

Fair share is a rather beautiful concept if you ask me. Much like permaculture, it goes against the societal norms. The thing is, in contemporary Western society, many people are frequently subjected to advertisements that tell them what is wrong with them and how they don't have enough.

People believe they need to buy constantly in order to fill the void inside of themselves with material possessions, such as larger televisions, finer cars, or creams to correct various "imperfections". All of those things may make you happy momentarily, but none of them will provide you lasting joy. Fix one "imperfection" and advertisements will alert you to another one. TVs fail, automobiles age, and the newest fashion goes out of style. You end up trapped in the dreaded rat race like everyone else and everything gets overwhelming and unpleasant.

Fair Share encourages us to leave the rat race behind. Instead of searching for the next best thing that will only bring you temporary joy, start focusing on what is currently available all around you. You'll begin to see that the reason we're all here isn't to maximize our individual wealth but rather to come together as a community after you stop stressing about all the stuff you don't truly need to acquire. Go outside again, find joy in all the little things that were there all along. Enjoy the simple tranquility of being in nature.

Going back to nature and doing it sustainably not only restores the soil but also provides you with more delicious, bountiful, and nutrient-dense harvests with less effort. There is no need to hoard it all yourself when you have such a wealth of resources right outside your door; you'll have in surplus. There's no way you could devour it all yourself, so you'll want to share your crops with others.

Additionally, when you help someone else, a lovely cycle of reciprocity begins, which keeps going. This is not about self-sufficiency, as I stated previously; rather, it is about self-reliance because you cannot give to others if you don't take care of yourself. Make the most of what you have and spread the abundance that permaculture farming will provide you.

If we all work together, we shall each receive our fair share. No one needs to be in need. Together, we can ensure that everyone has enough and that no one in our community is left behind.

PRINCIPLES OF PERMACULTURE

The permaculture principles are more akin to the laws of the game, somewhat of a guideline in this new path you've chosen. You must come to that conclusion for yourself; they are not the solution.

The evolution of the principles

Depending on your source, the initial brief list of principles has grown into a variety of lists. But Holmgren published a revised list of twelve guiding principles in 2002.

Holmgren's 12 Principles of permaculture

- Observe and interact
- Catch and store energy
- Obtain a yield
- Apply self-regulation and accept feedback
- Produce no waste
- Use renewable resources and services
- Design from pattern to detail
- Integrate rather than segregate
- Use small and slow solutions
- Use and value diversity
- Use edges and value the marginal
- Creatively use and respond to change

Let's take a closer look at what I believe are the four most important principles (especially to a new convert)

Observe and interact

It is important to invest a great deal of time analyzing your grounds. It gives you a better understanding of what you will face when designing and constructing the layout.

Capture and store energy

A wonderful illustration of this idea is the Scrumping Project. It takes a lot of energy to produce fruit, and if the fruit isn't consumed, that energy is wasted. By gathering any fruit that is going to waste, the Scrumping Project will "capture" that energy and "store" it by turning it into products like juices, jellies, and chutneys.

Creatively use and respond to change.

The Brighton Permaculture Trust has been continuously changing since its establishment in 2000. Since the organization's establishment, the key founders have remained somewhat active while also welcoming new members. According to the difficulties and resources at hand, the initiatives have altered. As a result, the organization has responded by coming up with creative solutions that adapt to the changes in funding.

Integrate rather than segregate

One thing I have been able to understand from my permaculture experience is that it is not wise to do it all alone. Of course, I do the physical work because It is my garden, but welcoming the ideas and opinions of others have helped me greatly in managing my space.

COMMONLY ASKED QUESTIONS

While I feel like I covered everything, in my experience, there are four areas in which a lot of newcomers get confused. Allow me to answer those questions before you ask!

Question

What types of activities do you wish to undertake in your garden? Will you use your yard to relax, prepare meals, play games, or perhaps even do some gardening yourself?

Answer

You need to think about your landscape's function in your daily life and the spaces you're ready to enjoy, whether it's hosting dinner parties on the weekends, having your kids play outside, or drinking coffee on the patio in the morning.

Question

Who will be using the area?

Answer

Consider the family members, young children, visitors, and pets that will frequent the garden on a regular basis. Many of your landscaping choices will be influenced by kids and pets. It will be crucial to consider who will be using your landscape so that you can take this into account when planning places and choosing the appropriate plant material.

Question

What style do you favor? Would you like a garden that was more informal or formal in style?

Answer

To achieve a cohesive look, think about what garden style would go with the design of your home. Additionally, it's crucial to consider all seasons in your climate. Later in this book, we will talk about how to choose plants that will produce interest throughout the year.

Question

How much maintenance will this require?

Answer

Whether you want to manage your garden yourself or hire a landscape maintenance firm, think about how much time you want to spend on it. It's important for you to choose the plants that will best suit your long-term maintenance goals, and we will talk about how to do that later in the book. Remember that plants are just one component of the design process, but they do help to create a sense of place. They will increase your sense of comfort in your surroundings and encourage you to go outside even when the weather isn't ideal.

I hope this chapter has broadened your general view of permaculture as a whole. You've found that it is a lot more interesting than maybe you initially thought.

In the next chapter, we begin our 7-step guide to starting up and successfully running your own permaculture garden. So if you already have a space in mind to use, it is time you two get very acquainted.

2

STEP 1: FAMILIARIZE YOURSELF WITH YOUR SURROUNDINGS

SCALES OF SYSTEM

Permaculture is not limited to your backyard or front lawn; you can actually practice it at a much broader scale. If you're not familiar with this aspect of agriculture, allow me to walk you through it.

Urban Permaculture

Okay, first, we have Urban Permaculture. Techniques at the urban scale are widely used in practice around the globe. We discover intensely farmed and extraordinarily fruitful systems on balconies, rooftops, little yards, and barren lots. Urban permaculture designs are densely populated, making the best use of the available land through intricate relationships between the flow of rainwater and sewage, food production, composting, sunlight, pollinator habitat, social areas, and the urban waste stream, which can provide free and inexpensive building materials for soil and structures.

Sub-Urban Permaculture

There are many opportunities in suburban-scaled permaculture systems. In fact, suburban sprawl benefits us, as permaculture co-founder David Holmgren put it. Detached homes are simple to convert, and the area around them provides access to solar energy and room for food production. Our spoiled decorative gardens already have a water supply, fertile soils, and easy access to nutrients. Numerous examples of suburban areas have been transformed into effective permaculture systems when we search the globe. Due to the reduced population density in the suburbs, there is a lot of room for larger gardens, livestock, tree crops, and land-based livelihoods.

Permaculture In the Public Space

Particularly in metropolitan locations, we discover a wealth of permaculture systems that benefit from the diversity of the "human ecology" present there. A community's sustainability and resilience are boosted via gathering areas, public parks, and communal food forests. A community can be resilient not simply because of its access to food and water but also because of its inhabitants' strong social ties, which foster a sense of belonging and reciprocal assistance. When we look to provide people with more opportunities to engage with one another, permaculture in public spaces makes excellent gathering spots.

Multi-unit Urban Development

Developers, planners, and landscape architects have used permaculture design to create multi-unit housing projects. With an emphasis on solar access, energy systems, water flow, paths, and a fertile landscape for humans and wildlife, permaculture involves integrating many patterns into the design of a housing subdivision. It is frequently only a matter of having proper positioning, orientation, and inter-connection between these elements that create the "main-frame" for a more regenerative settlement because development projects of this scale are already investing in buildings, utilities, drainage, and road infrastructure.

Ecovillage

Ecovillage, which is short for ecological village, is a collo-quial term for a permaculture-based village or community. Inspiring instances of how permaculture design is used to build intentional communities, help existing villages in traditional cultures, and enhance modern settlements can be found all over the world. Ecovillage design is a very rich field where the components of food, water, energy, materials, ecology, housing, and forestry are woven in with community political structures, economics, urban planning, and all the difficulties and potential of resource sharing.

Retreat and Healing Center

When building healing or retreat centers, several people use permaculture design. Both newly constructed retreat facilities in colonial areas and indigenous individuals using traditional medicine in their home countries have used it. Given that caring for the Earth and caring for people are two of permaculture's core values, the design philosophy is ideal for the therapeutic environments found in hot springs, yoga studios, ashrams, retreat centers, alternative health care facilities, centers for indigenous medicine, and eco-resorts. Permaculture's organic patterns can add a more refined aesthetic that is both useful and fruitful to these situations.

Homestead

The homestead scale for the purposes of this list denotes a rural property that is larger than a suburban lot but smaller than a functioning farm. This indicates that the managed area is between 0.5 and 5 acres (or 0.2-2 hectares). A homestead typically has more varied and comprehensive production systems for income and nourishment since it is large enough to support them. This is a highly common scale for permaculture projects worldwide since it allows for extensive animal rotation, the growth of big trees, ponds with more significant water collection, many larger structures, and successful cottage industries with many occupants.

Farms

In order to create successful companies and wholesome landscapes, permaculture farms are varied agricultural systems with a variety of various operations. In order to maximize output with the least amount of work as well as develop soil, improve the water cycle, and give farmers long-term resilience and abundance, permaculture farms are laid out in accordance with the shape of the land. Because they use natural patterns and combine people, animals, plants, trees, water storage, renewable energy, and other resources in a peaceful and beneficial way, permaculture farms differ from conventional farms in appearance.

Education Facilities

There are many overlaps on these systems scales, particularly for educational facilities, as many are also farms, homesteads, suburban lots, urban lots, or retreat centers. There is another category of permaculture sites called demonstration sites that are used to connect people with the systems, help them learn, and inspire them. Due to the fact that they are created to encourage public participation and education, demonstration sites frequently rank among the most extensive and diversified. If you wish to learn more about this fascinating area, a quick search online will show you locations of demonstration sites all over the world.

AGROFORESTRY ON A LARGE SCALE

Permaculture has been adopted by more than simply farms and homesteads. A few significant industrial-scale and commercial producers have realized the potential of perma- culture systems. They have made investments in significant projects that are enhancing the ecosystem, hydrology, and food security of entire regions. Corporations like M&M'S Mars Inc. in Vietnam have invested in these initiatives. Agroforestry initiatives share many of the same design ideas, methodologies, and strategies as projects that go by the moniker "permaculture". The possibility for industries to integrate permaculture design into current supply chains is expanding in this region, which would transform consumer spending into agents of land regeneration.

International Development

The use of permaculture design is also suitable for the field of international development. There are already numerous avenues through which funds, professionals, and other resources are contributing to the economic growth of other nations. Permaculture has its own initiatives to make global development sustainable and regenerating for economies, ecosystems, and communities. In several nations, permacul- turalists are already employed by numerous non-profit orga- nizations to lead training sessions and development programs.

International aid and disaster relief

Permaculture "first responder" teams have frequently been called to places that have experienced significant natural or man-made disasters. This frequently occurs with a non-governmental organization's assistance. In refugee camps and other areas that require reconstruction, permaculture is ideally suited to setting up secure and ecologically sound survival systems. In many disaster situations, appropriate technology and permaculture design can really help with things like the secure processing of human waste through composting, helping to keep the water clean and soils in place through erosion control and layout of camps, using appropriate technology for fuel-efficient cooking, and fast-growing gardens that turn trash into compost and food.

NATIVE PLANTS

There are plants that would look great in any permaculture garden or farm, but they may not always be the healthiest options for you or the environment. Garden centers may sell non-native or invasive species that are difficult to maintain or inhibit other plants' growth. Frequently, people who buy these are unaware of the environmental damage they might cause. Consider introducing native species if you've been finding it difficult to manage your garden or if you've been spending more time on it than you'd like. These are plants that have adapted to a place on their own without the help of humans. Here are several good reasons to expand your garden's native species.

Why you should add more native species to your garden

Foreign specimens are attractive and promising, but native species have their perks.

They Develop A Habitat For Wildlife

Native plants draw animals, increasing the biodiversity of the environment. When plants flourish in their natural habitat, such as your garden or backyard, the ecosystem and other plants benefit; when native species are growing, pollinators like birds and bees will frequent your property. For a great harvest, they'll assist in pollinating your fruit trees. Furthermore, the soil beneath these plants provides a habitat for tiny creatures and other living things.

The Environment Benefits From Them

Native species are better for the ecosystem since they can survive on their own without chemicals, in contrast to invading species that frequently depend on fertilizers and pesticides to grow. Additionally, native plants improve the soil's structure, which reduces water runoff and erosion. In addition, they can store extra carbon dioxide, which would improve the air quality surrounding your house.

They Conserve Water

Water conservation is among the most important issues relating to sustainability. Many individuals regularly misuse or waste fresh water for the sake of watering their gardens and yards. A significant portion of water use is for irrigation of lawns. Because they are unfamiliar with their surroundings, non-native plants need extra water.

They Offer Protection Against Invasive Species

Native plants in your garden can repel invasive species. Native plants have deeper root systems, are tougher, and are more resilient. It wouldn't be able to survive against native plants if an invasive plant species made its way into your yard. They are fiercely competitive and won't let undesirable plants take over your garden and yard. But to help them develop to be strong, make sure you maintain their care in the first several years after planting.

They Supply Food

There are numerous edible native plant species. You get all the advantages listed above and more when you plant a garden with native plants. The United States is home to a variety of native plants, including blueberries, cranberries, raspberries, and wild grapes. The ecology will benefit as you harvest more food from your garden. The less distance your food has to travel, the less carbon dioxide is emitted.

They Boost Soil Quality

You want to provide your plants with the greatest soil you can give them as a gardener. Healthy, nourishing soil is the foundation of a flourishing, fruitful garden. The nitrogen in the air can be absorbed by a variety of nitrogen-fixing native plants, which then release it as nutrients into the soil. Additionally, because native plants are more resilient, their deep-dwelling root systems have the ability to break up clay soils and let water through. They also help the soil retain nutrients and water.

How to find native plants near me

It's time to choose plants to cultivate now that you understand the fundamentals of native plants! Here are my top five resources for finding out more about native plants and deciding which ones to add to your yard:

Purchase a book on indigenous plants.

Although there is a plethora of knowledge available online, nothing quite compares to leafing through a book, especially one with fantastic plant photographs.

Look it up on the NWF plant finder.

You can use a fantastic tool created by the National Wildlife Federation (NWF) to locate the natural plants in your region that act as host plants. Simply input your zip code, and the site will display results ranked by the number of species of moths and butterflies that use those locations as host plants for their caterpillars.

Your yard will come alive and attract more insects if you give these kinds of plants top priority. It is a fantastic place to begin your search.

Investigate the Wildflower Center database.

The Lady Bird Johnson Wildflower Center keeps a complete database of native plants in North America. A list of flora in your state or Canadian Province can be found. When you have a list of plants, you may filter the results based on things like how much sunlight they need, when they bloom, how big the mature plant will be, and more!

Join your regional native plant society.

Joining a native plant society in your area is a fantastic opportunity to learn about local flora and connect with other native plant enthusiasts! Numerous organizations provide free classes on native plants online or in person. The Native Plant Society of Texas even provides the Native Landscape Certification Program. Native plant organizations frequently organize native plant sales and provide volunteer opportunities in addition to being a valuable source of information.

Locate a nearby nursery for native plants.

It's time to go observe the plants in person! The big box nurseries like Home Depot and Lowes are best avoided while looking for native species. Search instead for a nearby nursery that sells or specializes in native plants.

Check out the search results when you Google "native plant nurseries near me"!

SECTORS AND ASPECT: ORIENTATION, ELEVATION, AND SLOPE

It's always best you draw out a plan for the area, regardless of whether you are working from scratch or enhancing an existing location. In this drawing you will include the position of the boundaries, any trees, and other permanent features. You must also start gathering information about various habitats, such as areas of full sun or shade and the extent of any slope. You also need information on the water to identify rain shadows and free-draining or boggy areas, the direction of the prevailing wind in various seasons, what wildlife visitors come in and whether they stay, and so on.

All of these have an impact on how people use the site and influence how we successfully choose the layout and style. Simple and prolonged observation is a great source of a lot of this knowledge. If you have the luxury of time and patience, it is great to collect habitat data throughout the cycle of the four seasons. However, there is a lot of pre-work that we can complete that starts to fill in the blanks and create a dynamic picture of our site.

Sectoral Analysis

Using the straightforward tool of sectoral analysis, we can begin examining how the components of nature impact our site. We'll focus on the two major elements; Sunlight and Wind.

Sunlight

One apparent element is sunlight, for instance, as its location on our site and the time of day provide us with important details about how and where to position objects. The East is where the sun rises, due South is where it is at its highest (if you live in the southern hemisphere, it is highest in the North), and West is where it sets. In comparison to summer, the sun rises later in the winter, sets earlier, is lower at midday, and is not as far west in the winter. As a result, there are two distinct sectors for summer and winter sun. Our plants will grow most productively in full sunlight, but we should also learn to use our less bright spaces for plants that like shade.

The way the light hits our site is affected by hedges, walls, and buildings. In the summer, a hedge might only provide shade for a border in the early morning and late afternoon. The border, however, may be permanently shaded during the winter months because the sun is lower and the hedge is tall enough to block it.

Wind

Another element to consider is wind. There will be seasons when it might be consistently windy, which will indicate the direction of the wind that will be blowing where our site will be. The season and direction of the wind are crucial factors in how our site's components are laid out because the wind will always have an impact on plant growth by setting it back. Windbreaks could be necessary to add. The wind is channeled up and over solid structures, producing eddies that cause more damage. As a result, solid structures constitute the least effective windbreaks. Windbreaks that filter the wind are significantly more protective (such as a hedge or windbreak netting), and their impact can be felt up to 10 times as far away as the windbreak itself.

Frost and wind are related concepts. Warm air is lighter than cold air and has a tendency to fall downhill. Windy locations have this cold air pushed away, which lessens the possibility of plant damage. In terms of design, frost pockets typically occur in places where colder, denser air falls and gathers undisturbed. If possible, the wind should be directed into these pockets, or any potential windbreaks should be eliminated or reduced.

A sector distribution can also include rainfall and waterways. Buildings, bushes, and trees may cast water shadows on our sites. On the other hand, there will be places where any rain will constantly fall on the soil, and its subsequent drainage becomes important. Water and rainfall should always be used to their fullest potential, and if a problem (such as flooding or bogs) is associated with them, it should be turned into an advantage. Tree planting significantly affects water run-off, and water can be siphoned through a site by diversion ditches either to space out or to store off-site.

Another great thing about wind is that, together with water, you can create a conveyance that will move light and heat away from one place to another place that needs it more. By impeding the amount of wind or water that can flow to one place, it must go somewhere else. Heat and available light are easily transmitted throughout our system by air, water, or substances mixed with them. Particles or molecules carried in the air significantly impact these two variables.

The wind is one of these factors that we have the least control over in terms of storage or generation, but we may influence its behavior on site by excluding, reducing, or increasing its force and doing so with the help of windbreaks and wind funnels.

Winds under 5 miles per hour (8 km/h) pose no threat to crops. A speed of 15 miles per hour (24 km/h) reduces agricultural production and results in animal weight loss. However, a speed of 20–25 miles per hour (32–40 km/h) causes more mechanical harm to plants than any other consequence; in fact, in that strong of wind I have seen zucchini uproot and bowl around like tumbleweed.

Elevation And Slope

Few of us have the benefit of a level, completely flat site at sea level. In addition to having equally distributed sunlight, rainfall, and drainage, this ideal location would also have a temperate climate and may not even be windswept. Our dream must come to an end because, in reality, we all likely reside on hillsides with hummocks and bumps, and the weather dominates our conversations. We shall notice that the estimation of our site's slope and elevation complete the sector analysis.

Since the majority of us live in a semi-upland location, elevation is important. The growing season gets shorter as we ascend. The season's average temperatures are also substantially lower. Furthermore, insects may not thrive higher up, and they could not be present to perform the pollination tasks that some vegetables require, such as runner beans, which are rarely successful beyond 1000 ft (300 m).

If the slope is not too severe, it may even be advantageous. The land is exposed to the sun on south-facing slopes (north-facing in the southern hemisphere), which means fewer buildings or plants will cast shade. We get character and variation from changes in slope because they frequently mark the boundary between various ecologies. It is not surprising that human habitation can frequently be found between a forest and a plain, a plain and a marsh, or land and an estuary.

Regarding rainfall and waterways, the slope is also crucial. Water doesn't run down a slope; instead, it pours over the top until it reaches a sump. This means that the water doesn't stay at the top of the slope; rather it flows right down to the slump. Thus, a slope's top may be quite dry while its base is muddy. Better water management techniques (such as woody plants and swales) will keep the water as high up the slope as feasible or cause it to soak into the ground where it falls.

Aspect connects slope and elevation. It would be preferable to have a south-facing slope on a hillside rather than a north-facing slope on a valley bottom because the valley's cooler environment might not make up for the lack of sunlight. However, we must consider the positives. For example, a hill's shady side may prevent frozen plants from thawing too soon, reducing frost damage.

QUESTIONS TO ASK YOURSELF

Before you embark on this journey, take some time to consider a few things. First, ask yourself what kind of area do you have? Are you growing plants on your balcony or in the backyard? Is there any available rental spaces or community gardens where you may start your garden?

The ecosystem in your area is the next thing you should think about. Which local insects and plants are there? What native fruits and vegetables grow there, and when do they ripen? Since the idea is to preserve as much of the natural world as you can, before planting non-native species, you should think about what already exists there.

Following your research, you can design and construct your garden beds based on your available space and your projected crop size. If you need planters but don't want to purchase full-sized wooden plant boxes, you can instead purchase classic pots and planters from a gardening or home improvement store. Additionally, ensure you have harvesting tools such as shears, gloves, and other standard gardening necessities.

Finally, think about the environmental impact before doing anything in your garden. Research the impact of your agricultural endeavors on the local soil, animal population, and native plant species before making any big changes to the environment because permaculture is about treating the Earth with care.

Permaculture, as I mentioned earlier, means permanent agriculture. One thing will be constant throughout your permaculture journey, and that is the land. If you miss it here, it's going to be hard to bounce back. Luckily for you, you will practice what you read in this chapter and make the best out of your surroundings!

Creating the perfect space for your practice will pay off bigtime, but only with the proper maintenance. You'll need the most suitable plants and animals for your farming space, and I'm going to walk you through how to make those decisions!

3

STEP 2: CHOOSE YOUR PLANTS & ANIMALS

Have you ever heard of Findhorn Ecovillage? If you haven't, it's definitely another successful permaculture endeavor you should check out. The community is located on Scotland's west coast, and it combines an educational facility, an ecovillage, and an experiment in conscious living.

The Foundation takes pride in being sustainable in terms of social, economic, and spiritual factors in addition to environmental ones.

You'll get your own success story, too. But first, you must perfect your farm with the right plant and animal species. This chapter walks you through everything you need to know to achieve that!

EACH ELEMENT PERFORMS MANY FUNCTIONS

Each piece or component of a permaculture design is chosen and positioned to perform as many roles as possible in order to maximize a design's efficiency. We can only accomplish this when we are fully aware of all the characteristics of an element. When this element is a plant or an animal, we must have in-depth knowledge of the creature in question. This covers the conditions it performs best in and the range it can withstand, among other things.

A "functional analysis" can be used to pinpoint an element's needs, products, behaviors, and intrinsic qualities.

To pick suitable plants for your garden, you must consider the main characteristics of each species. Factors such as:

- **Form**

Lifestyle and form are all important factors (is it a ground cover, shrub, tree or vine, and how tall does it grow).

- **Tolerances**

What are the light requirements (shade, cover, or full sun), habitat needs (dry, moist, or wet, low or high elevation), climatic needs (arid, temperate, subtropical, or tropical), soil type needs (sandy, silty, clay, loamy, peaty, or chalky soil types), and soil pH needs (acid, neutral, or alkaline soil)?

- **Uses**

Elements may be edible, therapeutic, animal fodder, soil enhancers (nitrogen fixing, cover crop, green manure), or for site protection.

Here's how to apply the functional analysis using a willow tree as an example:

- **Form**

A weeping deciduous tree that can grow to a height of 75 feet (23 meters) is quickly propagated from cuttings.

- **Tolerances**

It prefers moist, well-draining soil and is frequently found next to streams and in other damp regions. It requires full light. It can tolerate salt and a variety of pHs and soil types.

- **Uses**

Source of Medicine: Salicin, an aspirin-like molecule found in the bark of white willow, is used to treat inflammation, pain, and headaches.

Source of Building Materials: Wood is used to manufacture furniture, tool handles, wood veneers, and toys, among other things. Willows produce the wicker used to weave wicker baskets and create fish traps. It can also be used as a source of fiber to make rope, string, and paper. Only willow is used to make the charcoal that artists utilize.

Form, Tolerance, and Use

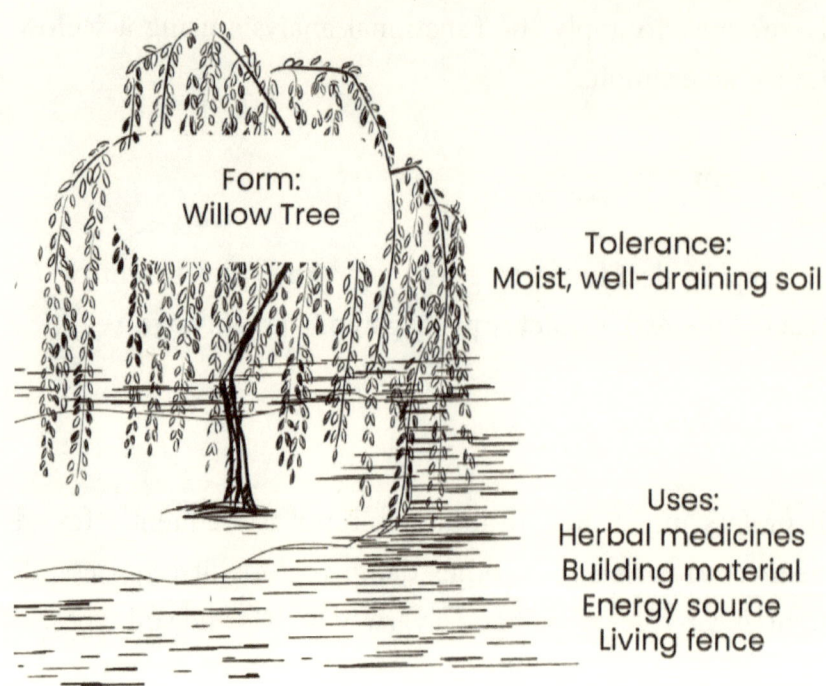

Form:
Willow Tree

Tolerance:
Moist, well-draining soil

Uses:
Herbal medicines
Building material
Energy source
Living fence

Energy Source: Willow is cultivated for its biomass, a sustainable energy source that lessens the need for fossil fuels and petroleum-based goods.

Ecological and Environmental Applications: They are applicable in the following fields:

- Riparian buffers: Organic barriers that keep chemicals out of lakes, ponds, and streams.
- Phytoremediation: Willows remove poisons from polluted areas.

Wastewater Management (Biofiltration): Willows are useful in ecological wastewater treatment systems because they can remove pollutants from wastewater.

Willows are frequently utilized for soil building, streambank stabilization (bioengineering), slope stabilization, soil erosion control, shelterbelt and windbreak construction, and land reclamation.

Willows are utilized to create wetlands and wildlife habitats during environmental repair.

Willows are used to build hedges, "living fences," and other living garden constructions and for regular landscaping.

- Living snow fences - Willows placed in strategic locations catch drifting snow.
- Agriculture - Farmers can utilize willows as animal fodder to feed their livestock.

Willow bark provides natural plant growth hormones that can be utilized to root fresh cuttings in horticulture.

CHOOSING THE RIGHT PLANTS

It might be challenging to choose the proper plants for your needs at times. Which tree species belong in your canopy layer? Which plant should you use to fix nitrogen? Which shrubs will work best for your location?

Of course, the typical response is, "it depends." You'll need to be more precise in your response and have a basic knowledge of the site's soil quality, microclimate, and climate to be effective. Even while this is only one piece of the jigsaw, choosing the best plants to work with is actually another.

I suggest you check out the Plants for the Future Database (PFAF), a fantastic online tool that, in my opinion, doesn't receive anything close to the attention it deserves. Rather than Googling, checking the many forums, and then becoming lost in the unending research, start with PFAF.

PFAF is a superb free online plant database that provides details on over 7,000 edible and other valuable plants, including their origins, edibility, and therapeutic applications. This extensive directory is a great resource for anyone who loves plants and wants to locate potential plants to cultivate. Furthermore, there are so many options for searching for information that it is almost overwhelming, including Latin or common name, plant family, environment, or use. Numerous helpful details on each plant are provided.

Take into account the qualities of the plants you'll need, your site conditions, and do a search to see how fantastic it is.

Let's look at one instance. Say you live in zone 7, according to the USDA Hardiness Zone Map, and you need a shrub that fixes nitrogen. You also know that your soil is alkaline. Your climate also has a tendency toward droughts, so your plants would need to be drought-tolerant.

What would be the best method for finding this kind of shrub? Here is a step-by-step breakdown of the procedure:

Open the PFAF homepage at pfaf.org.

I understand that at first glance, the sheer volume of information and options available can seem overwhelming, but for now, let's stick to the fundamentals. At this point, it is doubtful that you are aware of every property that is significant for a food forest guild, making it difficult and time-consuming to search based on them all.

Check the necessary boxes.

So, in our hypothetical example, we have a general notion of what we require: a shrub, hardiness zone 7, nitrogen fixation, with tolerance to droughts and alkaline soils. Accordingly, we will look for these kinds of plants and check all the essential boxes. Do yourself a favor, and for the time being, put all the other properties on the back burner.

Perform a search and display the results.

When you conduct a search, a page based on your inquiry will be displayed, along with a numbered list of plants that might be appropriate for your particular circumstances.

The Siberian peashrub, which Geoff Lawton likes using, the Russian olive, which Martin Crawford enjoys, the goumi or the silverberry, which Michael Judd enjoys, the autumn olive and the sea buckthorn, which Ben Falk prefers, are just a few of the intriguing plants we have available.

Make a list and select the most suitable plants.

Finally, click on each plant's name to learn more about it and its traits. You should read up on each plant and how it might fit into your guild, so doing this is really worthwhile.

Now that you've copied the data you discovered into a blank spreadsheet, you have a list of nitrogen-fixing shrubs that may thrive in your climate and can withstand sporadic droughts.

After doing this, you can simply repeat the process for any more plant functions that you can think of that you'll require in your food forest guild to create a master list of plants.

PFAF is my go-to site for discovering new plants, but I understand it can be difficult to navigate. There are suitable alternatives that are easier to use and offer equally useful info. You can try any of:

- Lady Bird Johnson Wildflower Center, www.wildflower.org
- New England Wildlife Center, www.newildlife.org
- Missouri Botanical Garden, www.missouribotanicalgarden.org
- The Ferns Website, www.theferns.info

BEST PERMACULTURE PLANTS FOR BEGINNERS

This section will concentrate on permaculture plants that flourish in mild to chilly temperate regions, such as those found throughout most of North America and Europe.

Several common traits that all plants share make them excellent choices for use in permaculture gardens, regardless of the environment where you live. Allow me to walk you through it.

You want to pick a plant that:

Requires minimal upkeep

A permaculture garden is one that (most of the time) looks after itself. Therefore, choose plant species that can survive even in less-than-ideal situations like poor soil or hard weather.

Has deep roots

Deeply-rooted plants are a permaculture gardener's best friend, especially if you're planting in a location with inadequate soil. They will bring nutrients from deeper soil to the surface, where they may be obtained more easily.

Belongs to the family of legumes

Due to their ability to capture nitrogen from the air for other plants, legumes are frequently used in permaculture gardens.

Produces loads of foliage

In a permaculture garden, plants with lots of leaves make excellent mulch. Certain plants function as living mulch to keep the soil moist and keep weeds from getting sunlight. Others are cut down and used as mulch, which is placed back on the ground where it is needed.

My Top 16 Permaculture Plants

In my experience, I've been able to observe over a dozen plants that do a little better than others. Not that you are constrained to a small variety of plants while designing your garden. There are many options to choose from, but here are my top picks.

1. Comfrey

Many farmers and gardeners see comfrey as a weed, just like many of the best permaculture plants. It may be considered a weed, but it's a weed with loads of benefits. For instance, comfrey has lovely bell-shaped flowers in shades of blue, pink, purple, or white that attract pollinators.

It gives refuge to beneficial insects beneath enormous leaves, assisting in pest control in the garden. You can also use their rapidly expanding leaves as green mulch to fertilize the soil.

2. Hazelnuts

Hazelnuts can develop into tiny trees or bushes. They make excellent windbreaks on their own and combine with larger trees, like apples, to form an excellent permaculture guild.

These trees can withstand shadow and act as a neutral buffer between plants that don't get along with one another. The high-calorie nuts they produce are plentiful and can be consumed raw, roasted, or used to make flour or oil.

Hazelnuts are an excellent addition to your food forest because they can live for up to 50 years.

3. Jerusalem Artichoke (aka Sunchokes)

Due to their extreme hardiness, Jerusalem artichokes are one of my favorite permaculture crops. Apart from what they naturally receive from rain, they don't need any upkeep or water.

In spite of their name, they are neither from Jerusalem nor are they artichokes! They are actually a kind of sunflower that is indigenous to North America. They develop into a sunflower-like height and have a charming small yellow blossom on top. Additionally, sunchokes yield wonderful edible tubers that resemble ginger and store nicely for the winter.

Jerusalem artichokes' tall stems are sturdy enough to be used as a trellis for beans and other climbing plants. They even function somewhat as a windbreak.

Growing these annuals in an area where you don't mind them encroaching is advisable. Since the tubers are so hardy, they can spread quickly to create extensive swaths of the plant.

4. Fiddleheads

Once cooked, the fern fiddleheads have a flavor that is quite similar to asparagus. They are only edible briefly in the spring before they begin to unfold. However, as you wait for the rest of your crops to develop, they provide a fantastic supply of vitamins and nutrients.

Fiddleheads are typically thought of as being foraged from the wild. But there's no reason why you can't incorporate them into your personal food forest. They will flourish if you put them in a moist, shaded area.

5. Mulberry

Here's another little tree that yields a lot of nutrient-rich vegetables.

Since they lose their freshness quickly after being picked and are readily harmed in transit, mulberries have never been very popular. However, that doesn't mean they aren't tasty! They taste delicious eaten straight from the tree or used to make jams and other types of preserves.

Mulberry trees make excellent homes for birds and other wildlife because they grow quickly and are incredibly sturdy. They can be used as mulch, and their big leaves provide a lot of shade.

Mulberry stains easily, so put your trees far from your home to prevent a mess once the fruit starts to fall.

6. Arrowhead

Despite the fact that most of the permaculture plants on this list are land-based, I had to include one of my favorite water vegetables in case your property has a pond.

You should roll up your sleeves and get a little bit wet for these scrumptious arrowhead tubers! They have an earthy, nutty flavor and cook similarly to potatoes.

Give your frogs, turtles, and other pond-dwelling creatures a shaded area to hang out under these plants.

7. Mint

The only plant on this list that everybody is likely to identify is mint. It can be used in countless ways, including as a delightful herbal tea or in a sauce to serve with lamb.

Mint is nutrient-rich, effective at hiding foul breath, and even has indigestion-relieving properties.

You've undoubtedly previously heard how quickly mint can grow. This makes it an excellent living mulch to use as a base for higher plants.

Mint blooms are also a favorite of bees; however, the scent repels many undesirable bug species. Thus, everyone benefits.

8. Red Clover

Clover is excellent cover crop material. It helps to obstruct sunlight that dries up soil quickly and combats erosion.

It is also a nitrogen fixer, converting nitrogen from the air into nutrients that plants may utilize in the soil. Your soil will benefit much in the long run from this.

The majority of animals adore red clover, and it is perfectly edible. Therefore, it won't cause any problems for your cows, sheep, goats, or other livestock to graze there. Alternately, if you want to entice wild animals like deer onto your property, this is a food they like to eat. It is adored by bees as well, and it produces delicious honey.

9.Ramps (aka Wild Leeks)

Ramps are another crop that only appears briefly each spring, like fiddleheads. Ramps actually only sprout within a brief window of time after the snow melts and before the trees begin to grow new leaves.

Before they go dormant for the rest of the year, you have around six to eight weeks to harvest them. You can pickle them or turn them into a pesto to keep them around all year long.

Under deciduous trees, ramps flourish particularly well. They dislike being too hot, and without lots of shade, they can't make it through the summer. So, keep that in mind when designing your permaculture garden.

10. Asparagus

Although it takes time to grow, the common asparagus plant is an excellent permaculture crop. It will take two or three years after you first plant crowns or seeds before you can begin harvesting your first stalks. However, after that, your asparagus will grow stronger and bigger every year!

Asparagus makes a wonderful permaculture plant, considering how little upkeep it requires once it's established. Year after year, it will keep producing food. It also makes a fantastic addition to most gardens because it is resilient and adaptable to a wide range of conditions.

Asparagus is particularly beneficial as a companion plant for tomatoes. Nematodes, which can affect tomatoes, are plant parasites that are repelled by asparagus. On the other hand, tomatoes fend off asparagus beetles, which eat the plant's roots.

Asparagus can only be fully appreciated after being plucked from the garden. Asparagus from stores is typically fairly woody and dried out in comparison. It's worth creating a small patch for yourself if it's one of your favorite vegetables!

11. Flowers

Picking your favorite flowers to gaze at is advised; they are also delicious in salads and other foods!

Hibiscus, chrysanthemums, nasturtiums, violets, and magnolia are a few tasty flowers to think about planting.

You should add edible flowers to your permaculture garden for reasons other than aesthetic appeal.

They aid in attracting pollinators like bees and butterflies, for starters. Increased pollination increases the overall crop output of nearby fruits and vegetables. In addition, certain edible flowers have a perfume that keeps animals like deer and rodents away.

12. Stinging Nettle

Stinging nettles taste great and are filling when cooked. They can be consumed like steamed spinach or added to soup or tea. They contain significant amounts of nutrients, including vitamin C.

Nettles become edible and lose their sting when heated. You might, however, have a love/hate relationship with this plant while it is in the ground. You shouldn't have any issues collecting them as long as you use gloves and long sleeves.

Since it might help with inflammatory disorders like arthritis, some people even come to appreciate the sting. However, if there will be young children or pets on your property, you might want to skip this one or only put it in a place where they can't get to so they don't accidentally hurt themselves.

13. Strawberries

As long as you plant strawberries under the correct growing circumstances, they make for a tasty treat that returns every year. Additionally, unlike some other annual berries (like raspberries), you won't have to battle through thorns and brambles to get to them.

Strawberries are not only a delicious fruit but also spread out on the ground as a living mulch and aid in water retention on those hot, sunny days.

14. Walking Onions

If you enjoy onions, you might want to grow a lower-maintenance kind in your garden by planting walking onions. Other names for them include tree onions and top setting onions.

In essence, the plants get heavier due to the growth of bulbs on their tips, causing them to slant over and replant themselves.

They don't need much care to develop and spread themselves across your garden.

Allow this variety to take care of the labor if you're sick of growing onions from seed every year!

15. Black Locust

The black locust tree has many beneficial functions in the permaculture garden, even though it doesn't produce any edible fruit.

It's big enough to function admirably as a windbreak or hedgerow. Additionally, its wood creates strong, long-lasting timber, and its blossoms are great for honey bees and other helpful insects.

Additionally, black locust seed pods fix nitrogen, and the leaves make excellent mulch.

16. Hops

Without some sort of vine, a permaculture food forest is incomplete.

Some folks could favor grapes, beans, peas, or hardy kiwi kinds. But I'm a huge hops enthusiast.

Hops vines spread rapidly and offer a lot of shade. Bees can also find pollen in their blossoms.

Of course, hops' main purpose has always been beer production. You could certainly try brewing your own with these. However, dried hops can also be infused with lavender to create a soap or added to pillows to help people fall asleep.

Try to choose plants for your permaculture design that complement one another and need little weeding, watering, or fertilizer.

Remember to consider your personal growing circumstances while choosing the plants that will work best in your garden.

Select plants that enhance the soil, draw beneficial insects and wildlife, or help your other crops in a complementary manner. The best part is to try to include some permaculture plants that produce foods you like to eat!

Your entire permaculture garden doesn't need to be built at once. You can start by adding a few fresh plants to your backyard at a time. That way, you can enjoy the process, and you'll have a great permaculture garden planted in a few years.

ANIMALS AND PERMACULTURE

Now let's talk animals! The employment of animals in permaculture design is a crucial component that contributes to a system's variety and speeds up the system's progress toward maturity.

System components can be integrated by utilizing animals' innate propensity to perform beneficial work as part of their life function. This increases the complexity of interactions that are feasible and is consistent with ecological reality's natural tendencies. Animals like working and are excellent at producing goods like eggs, meat, milk, cloth, and skin that are useful to humans.

Your permaculture system will benefit from having a wide variety of helpful animals since they can fill numerous natural niches in an efficient way.

Here's what to consider when selecting your "permanimals."

Every animal needs food, drink, shelter, and room to be an individual. You should always consider how much time and resources their upkeep will demand, whether you're raising cows or tasty grasshoppers. Frequently, this entails delegating duties. Have a backup plan in place in case you become ill, whether it be a large family that divides tasks, a communally milked neighborhood cow, or if you want a vacation you may plan to hire someone to help with chores while you're gone.

You should think about the following before raising any animal:

Food

- What is necessary for the health of the animal?
- What goodies does it desire (so you can make sure it gets some, but not too much)?
- What crops and/or rotations will supply a large portion of their food?
- If you need to buy feed, where is your best place to do so?

Water

- How much do they require each day?
- Which delivery maintains water purity the best?
- If it freezes in the winter, what will you do?
- What spotless (metal) roof could you be able to collect water from?
- How much money or resources would you need to provide water for each paddock if your animals are housed in them?

Shelter

- How much space does each animal require for shelter?
- How do they prefer to be milked, sleep, or lay eggs?
- What maintains the animal's body temperature?
- Does the shelter move or remain in one place?
- Can your animal withstand being without shelter all the time?
- What does your animal enjoy doing in a space?
- Could it be free-ranged, paddocked, or mobile? Keep in mind that, despite the possibility of all animals being free-range, your site may have restrictions due to the requirement for animal labor or the animal's propensity to get into things it shouldn't.
- What kind of framework will keep them where you want them to stay?

- What kind of building will safeguard them from predators, such as vultures, foxes, or snakes?

Waste

- What kind of waste does the animal produce, and do you wish to use it?
- What environmental safeguards ought to be taken?
- How will the waste be handled?
- Is there a need for a sacrifice area or a water source only to treat manure and prevent muddy feet issues close to the shelter?
- What tools and supplies could you require for this sacrifice area?

Companionship

- What kind of companionship is required by your animal?
- Could it possibly be inter-species?
- What are the telltale indicators of unhealthy companionship (i.e., mistreatment of your hens by roosters)?

Guidelines for integrating animals into permaculture systems

It's a significant decision to raise livestock, but remember that you don't have to start breeding right away.

Possibilities for trying something out initially include:

- Purchase a few young animals to fatten and raise
- Examine several breeds.

And take your time! Before beginning to care for another type of animal, give yourself time to learn and build expertise with the first. Here is some specific advice for a few various types of typical farm animals.

Sheep

Which breed of sheep to choose when starting with sheep is one of the first choices you must make. Due to the fact that some sheep breeds are harder than others, this will also rely in part on the sort of terrain you have and the temperature. Mountain and hill sheep are tough creatures who can endure a variety of challenging environments all year long. Sheep in lowland areas require greater grazing and some cover. Your motivation for keeping sheep will also be a factor. The products of sheep include wool, meat, milk, and more sheep!

Although all sheep produce wool, some produce finer fibers than others. If producing wool is your primary motivation for keeping sheep, it is worthwhile to speak with spinners and weavers and to attend agricultural fairs, particularly if rare breeds of sheep are participating.

All sheep can produce meat, but some reach greater weights than others, and the taste of the meat from various breeds varies considerably. Just imagine who the meat is for. Is it just for your family? Do you intend to sell it instead?

For their lambs, all sheep produce milk. Several breeds can also be milked, usually for the purpose of manufacturing cheese. For instance, Manchego and Wensleydale cheeses are made from sheep's milk.

Your sheep will have to be sheared yearly for their comfort and health, even if producing wool isn't your main goal. However, because nylon is more widely used for clothing, blankets, and carpets, the value of commonplace fleeces is lower than the expense of shearing them. Instead of being recognized for its warmth, breathability, and lack of microplastic pollution on the environment, wool has thus turned into a waste product. Consider getting creative to turn this "waste product" back into something valuable. The wool can be spun into yarn and knitted into socks to be sold at local farmer's markets where there are customers who value limiting waste and prefer to support local.

Cows

Sheep require less space than cows, which are larger animals. However, they work well with sheep in a permaculture system since sheep consume shorter grass than cows, allowing sheep to follow cows in the rotation system.

Nowadays, there are separate varieties of cattle for dairy and beef, although the majority of the older types are dual-purpose animals. Due to the fact that a cow must give birth to a calf in order to provide milk, a dual-purpose breed makes sense if you are considering getting a house cow and you want a calf from which you may eventually obtain beef as well as deliciously fresh milk.

In order to keep the meadow's biodiversity and ensure that the soil can continue to retain carbon, pasture-raised cattle are a huge help.

Goats

Since goats browse rather than graze, they prefer eating trees and bushes to grass. They are very quick and can climb and jump, so effective fencing is essential.

If you don't have a sturdy, long-lasting enclosure for your pet goats, don't even try to maintain them in a food forest environment. A herd of goats can quickly destroy an orchard. However, that same goatish behavior becomes a positive when you use them to eliminate acres of poison oak and/or brambles, which they can do in just a few days. The distinction is in the design. Put your animals and plants in a position where their natural strengths are a benefit to you, and what could have been considered an annoyance is now a joy.

Chickens

There are many benefits to keeping chickens, and they don't simply pertain to eggs! Among many other things, they can be beneficial for gardens and make wonderful family pets.

Anyone who has had properly cared for, content hens will attest to the fact that they are remarkably kind and affectionate. We'll cover only a handful of the many benefits of keeping hens in this list.

Perhaps the biggest benefit of raising chickens is the availability of fresh eggs. With its abundance of vitamins, Omega-3 fatty acids, and beta carotene, eggs are incredibly healthy and nutritious. Even less saturated fat can be found in fresh eggs. Plus, you have the added benefit of knowing that the eggs you're eating came from chickens who are content and healthy.

Making compost is another convincing justification for keeping chickens. Your vegetation adores it! Every gardener appreciates quality compost, and chicken litter is one of the best fertilizers. If you keep your chickens in a coop, you can clean up when it needs changing by just dumping the bottom into the compost. And it is so much better for your garden than commercial synthetic fertilizers.

Alpacas

Alpacas don't damage your property because they only nibble at the grass. The pastures still need to be mowed. Both sexes often use the same places to urinate, though male alpacas can be a little neater than females. Therefore, regular scooping can keep your pastures attractive and lush.

Raising for meat doesn't have to be the first or even the last choice when it comes to alpacas. They are creatures of fiber.

Thus, we gladly enjoy the benefits of their annual shearing. Spinners, fiber cooperatives, and fiber mills can purchase their raw fiber.

Alpaca dung is an excellent soil conditioner. It doesn't need to be composted due to the alpacas' modified ruminant status and high nitrogen content. It can be spread directly on your garden beds or sold to neighbors who are looking for a healthy, natural fertilizer.

Rabbits

If you're thinking the only purpose for these perceptive, attractive, and amusing animals is ingesting, think twice.

Rabbits are kept on many permaculture sites for manure, meat, and weed control. They are simple, quiet, and take up little room, but it is difficult to keep them free-range because raptors rapidly steal them. If you have a problem with that, you wind up keeping them in cages, which may not be ideal for many designs.

Pets

When considering animals in a permaculture system and building our entire system to be sustainable, we must keep in mind that pets are a part of our life and our families.

Although I won't go into great detail here, know that pet waste can be composted, but it's not always recommended. There is generally a higher potential for pathogens in dog and cat feces, so special care needs to be taken to make sure all pathogens are killed off in the composting process.

Selecting the right plants and animals for your farm will perfectly complement and enhance its growth and development. With these guidelines, your garden will be teaming with all sorts of supportive life in no time, that is after you've designed a suitable garden layout.

Don't know what a garden layout is or how to construct one? Then let's head on to the next chapter, where I reveal the not-so-hidden secrets of Garden Layout.

4

STEP 3: DESIGN YOUR GARDEN LAYOUT

"**B**een meaning to post for a while to share part of our permaculture journey! My husband and I have about 17 acres in the woods with a stream in rural New York state. We have 15 chickens and got two goats this spring, and we've slowly added perennials and sheet mulched beds over the past two summers.

Also been catching rainwater and making compost tea with it regularly this summer, and it's made a wonderful difference in the overall health of our plants. Little by little, we are adding in, rearranging, planning, and replanning as we observe what works and what doesn't. For instance, we live on very hilly property, so I built some swales...by hand...and discovered we need to go way more hardcore to stop the erosion. However, since we don't have an excavator or anything, I added some hay bales below the swales after watching one of the lessons, and it's made a huge difference!"

- Katherin from Quora

Success stories such as these are only possible when the garden is handled with care and wisdom. Wisdom is what you'll need for the next step; wisdom and creativity.

HOW TO DESIGN YOUR GARDEN

Now that you have an understanding of the permaculture basics, you're eager to put this exciting new knowledge to use. You want to employ a permaculture design, but there is an issue; where do you start? How do you take this information and actually use it to design the layout of your garden?

Everyone is talking about the enigmatic permaculture design concept, but it's difficult to explain the fundamental methodology without attending a Permaculture Design Course. Permaculture design is undoubtedly a shady and mysterious type of alchemy. When I dove into this topic, I found books and encyclopedias available, but what I needed was a clear explanation with easy steps, something that wasn't too overwhelming so that I could get started.

What I found is that when we dissect an idealized permaculture design, we can observe four essential, connected actions:

- People Analysis and Assessment
- Site Analysis and Assessment

- Design Concept Development
- Implementation & Evaluation

I will outline the stages that lead to the final design and briefly discuss the implementation phase in this section.

Before we go any further, let me just say that permaculture design is a very in-depth topic that I could take up an entire book or even an entire course to teach. The purpose of this book is not to overwhelm you with too much information but to show you how to get started. It's much easier to start small, learn a little at a time, and add elements into your permaculture garden one at a time so you can learn and evaluate how they all work together on your property as well as enjoy the process along the way. Here is a general outline that will get you started on your journey of enjoying all that permaculture living has to offer.

Analysis and Assessment of You

You'll face numerous challenges along the way, so having a clear vision can help you accomplish your objectives.

Let's first think about the essentials before moving on. To do that, take some time to get to know YOU. What makes this so crucial? Be aware of who you are because you can rely on your strengths. Who are you, since this project begins with you? According to David Holmgren, "before I did anything for a customer, I wanted to know what kind of person they are; are they an animal, plant, technical, or people person?"

What personal assets do you have that you could leverage? Start by building on your abilities before you try to overcome your flaws. This is the best way to achieve success.

Identify Your Goals And Vision To Hasten Design And Implementation

Imagine the future—what do you want? Your project's vision serves as its motivation. This will serve as your north star, pointing you toward accomplishing your objectives. Eventually, when things get difficult, it will serve as a reminder of why all of this is important.

Next, state your objectives clearly. What are you trying to accomplish? What do you need your farm or property to provide for you—for example, what kinds of food, herbs, medicinal plants, firewood, timber, or other products would you like to have? What do you want your property to be—both in terms of how it would appear and function and what might occur there?

In their book *Edible Forest Gardens*, Eric Tonsmeier and Dave Jacke write that "articulating your answers to those questions is a core effort of the design process." The greatest approach to expressing your ultimate goals is in writing. By the time you're done, you ought to have a written statement outlining your goals.

To understand what you can provide, list your personal assets and limitations. How much time, resources, and effort can you commit to the landscape's planning, execution, and upkeep? Which of these resources could you be able to obtain through your friends, family, neighbors, or the local community?

Examine the resources available for the project. What tools, supplies, and resources do you have already? How much cash is available to invest in the project? Is it offered in a single payment or in installments over a longer period of time? Are there any options for external funding?

Additionally, take into account any possibly restricting personal characteristics; your health, age, social circumstances, or anything else you can think of may limit you in some manner.

Site Analysis and Assessment

Although every landscape is a whole system made up of several components, you should analyze them now. You'll need a map as your first item. Here's a step-by-step guide on how to accomplish that.

To have a foundation for your design, acquire a base map or create one.

The design is built on top of a base map. Google Earth, Google Maps, and other online applications of a similar nature make it simple to obtain maps. The next stage will benefit you greatly if you can obtain a contour map showing the topography.

If you can't get a map, you can still doodle one, making a basic sketch of the site that highlights its most important elements and what is already there. I would advise including names, the north arrow, the location, the scale, and anything else that cannot be modified. In order to build other layers of your design and any field observations, keep in mind that the map is the most crucial component.

Research and observe the site to get information about it.

▷ **In-person field observation**

Good observation is the key to permaculture. For about a year, you should ideally do nothing but become accustomed to the four seasons, the current weather, and environmental patterns. You should know where the wind is blowing, where the water is flowing, where pollution is drifting, where your neighbors are walking, and how much sun and wind you are exposed to.

Take surveys while strolling the area. Which animals are present? How does the soil look, and does it differ depending on the location? Which plants are present both on the property and in the wild? Determine any free or inexpensive resources that are present at the location or close by, as well as any potential or known sources of water. What are the site's boundaries? Take note of the paths and buildings that are already there. Keep a list of what you observe.

▷ Data derived from other off-site sources

After making the initial observations, Toby Hemenway advises conducting more studies online, with local experts, or in books to find out more about qualities that cannot be directly witnessed.

Although data from other sources can be useful, direct observation provides the majority of information. You can get more information about precipitation, hydrology, insolation, and wind speeds online.

Learn about the site's regional (geographic) and bioregional (flora and fauna) surroundings, as well as the history of the site and the neighborhood's development trends. To acquire priceless information that might not be available from any other source, speak with neighbors and other members of the community.

Evaluate The Site To Understand What You Have

The majority of the information you require has been gathered, and you are now prepared to analyze the data to determine what it tells you.

Eric and Dave write: "Analysis and assessment dissect the site so you may deeply comprehend each component." Now that you have organized your observations, you should explore every aspect of the landscape to determine the most significant aspects that you need to account for in your design. Start with the climate since it impacts your site most and cannot be changed. Examine the data on rainfall, solar radiation, the last time it frosts, and plant hardiness.

Next, draw the site's borders (site dimensions) on your base map and list the infrastructure already in place, including any buildings, roads, walkways, and fencing.

Examine geography, slope and aspect, and significant terrain features like peaks and valleys. Some examples of these are water drainage and watercourses, as well as creeks, dams, and ponds. Take note of trees that are already present, as well as the main flora growing there, the site's soil conditions (wet, dry, boggy), and types (clay, sand, gravel, rock).

On your map, Make a sector analysis of the external forces that are affecting the site, including the sun, wind, flooding, fire, pollution, and animals. Locating and outlining the distinct sectors will enable you to later position your design elements in suitable connection to the outside forces entering the site. This will influence your choice of plant species and location, allowing you to prepare for a wider variety of crops.

Summarizing your existing situation by writing your observations and drawing the land you have to work with will help you view the big picture and give your design a direction.

The Process of Design

Making connections between your vision and your observations is the essence of design.

You should decide what goes where in your area by utilizing design techniques and following the permaculture principles. Each approach offers a means of assisting you in making connections between various design components.

In his *Designers' Manual*, Bill Mollison lists a variety of design methodologies; if you want to take a deep dive into those, I suggest reading his chapter on design methodologies. But in this book, we'll keep it simpler, and I'll summarize just two of the most typical methods.

1. Start with the big picture, then fill in the details

Here is Dave and Eric's viewpoint: The primary goal is to create a rough layout with a focus on the relationships between the main features, elements, and functions, as well as approximations of sizes, shapes, and locations.

Draw rough bubble diagrams on your map to locate, shape, and size the necessary areas rather than the individual parts and pieces at the beginning. Use the data from the analysis and assessment phase to help you determine the rough layout and design connections in this phase.

It's best to start with infrastructure and the fundamentals: Structures (house, outbuildings, portable structures), access (farm roads, tracks, paths), and water systems (water storage, harvesting, irrigation). These are what Geoff Lawton refers to as "mainframe permaculture." After determining the infrastructure layout, determine main areas based on micro-climates and choose where to plant gardens, crops, orchards, and forests. If you are including animals in your permaculture farm, at this point you will want to mark where any fencing (permanent, living, electric) will be needed.

2. Make a thorough design and consider every last detail

While the first method was for the person who prefers to think in terms of the big picture, this method is for those who prefer thinking in terms of small details and precision. Using this design methodology, the drawings have sharper edges, are more exact, and have more clarity. "We need to think about how each piece of the design behaves and what its relationships are with the other pieces of the landscape and with us, the human inhabitants," Toby advises. He advises employing a linking process, commonly known as a "needs and yields analysis," wherein each plant, structure, or other element in a design should, in an ideal world, have its needs met by other design elements and provide yields that in turn nourish other elements.

If you're having trouble making connections, try the "random assembly" method of analysis, in which you make a list of the key components and investigate the results of randomly combining them. Unexpectedly beneficial connections can result from creative thinking. Random assembly aids in breaking through creative blocks and rewards us with connections and combinations we probably wouldn't have considered otherwise.

You will decide on a variety of design details and sketch the specifics of the different planting beds, trees, walls and fences, patios and decks, and any other design elements during the detailed design phase. Map out your water layer, roads, and buildings in detail. Determine the desired species for planting areas, pay close attention to your site preparation, and develop a plan for your strategy.

The ultimate goal should be to produce precise line drawings showing each component's size, shape, and location. Appendices like maps, drawings, plans, layouts, details, part lists, and photos are among the suggested attachments by Bill Mollison in his *Designers' Manual*. Create construction diagrams with notes, species and material lists, calculate the cost of various development stages and create a list of revenue-generating strategies.

Implementation and Evaluation Timeline – Make A Plan Of Action

Plan out what you're going to do at this point, define a timeframe, and create a "to do" list that you can refer to if you get stuck on a particular task. Put together a simple timeline with suggested chores. Make a strategy for what you will do in years 1, 2, and 3. Analyze your priorities and financial situation to decide how much you want to spend during the establishment phase.

It's best to start by considering the necessary infrastructure, and when you are ready to install permanent structures, you should begin at the entrance of your home and move outward. It makes sense to start by taking care of what you already have, restoring what you can, and adding new components to the system along the way.

Go back to check on the work you have done every week to see how it is coming along. Are different elements interacting the way you thought they would? Is the type of fencing you chose working to keep your animals where you want them? Have new trees that you planted changed how the sun and wind affect the plants around them? As you are seeing your design slowly come together, does that give you new ideas to adjust your plan going forward to make it even better?

Every component of design interacts with and feeds the others in an integrated process. After analyzing and evaluating yourself and your site, you integrate the findings of your investigation into a logical whole—a design concept. Furthermore, you must plan the components of that whole so that you can compile all the materials required to build it. The design must then be put into practice on the ground, and it must be frequently evaluated.

CHOOSING YOUR LOCATION

Think about where to put your vegetable garden before you dive in to see if you have a "green thumb." Here are some suggestions for selecting the best location for your backyard or neighborhood garden.

Exposure to the sun

Most veggies require about 8 hours of sunlight each day to grow well.

The majority of summer crops that bear fruit, such as tomatoes, peppers, cucumbers, zucchini, and squash, require a lot of light to grow, whereas root vegetables, such as carrots, beets, onions, and potatoes, can thrive with just 5 to 6 hours of daily sunlight. Lettuce, chard, arugula, and spinach can all grow with only 3–5 hours of direct sunlight.

There will therefore be some crops that require more care than others, but generally speaking, the more sun you can get, the more fruitful your garden will be. You should aim for a full day of sunlight throughout the growing season.

A south-facing slope (here it is again) on an open site free of any shadows from trees, structures, or buildings makes the ideal location for a garden. You benefit most from solar exposure, thanks to the south-facing hill. It receives more sunlight than the east, west, or north-facing slopes. The entire garden area also warms up faster since the sun's rays directly contact the soil surface.

Sunlight should not be blocked in any way if you want to fully benefit from it. But if you don't have a large open area without shade, you can still make the most of the available sunlight. Start by finding the shape and size of the area on your site that gets the most sun. If there are any spots that get a full 8 hours, plant your 8-hour sun-loving vegetables there. If it is less, choose vegetables from whichever list applies to you so that you can get the best yield possible. If you don't have any spots that receive 8 hours of sun, but you really have your heart set on tomatoes or peppers, try getting creative. Is there another area of your home that gets full sun but is not a typical garden location? Try planting your tomatoes in a container and placing them in that area.

Watering and Water Sources

Ideally, you will want to place your crops close to a clean water supply to make watering your garden or planting pots convenient and easy. Water your garden in the morning so leaves will dry quickly. When foliage stays wet for too long it encourages plant illnesses from fungi and bacteria that may harm plants.

Garden Soil Quality

Vegetable plants favor soils that drain well and do not pool after heavy rains. Additionally, pick locations where the soil is free of any potential contaminants, such as those close to sidewalks where ice-melting chemicals may have been applied or spots where water from road runoff may drain.

Distance from your home

The site's proximity to your home should also be taken into account. Your garden should be situated as close to your home as is practical, whether you live on a 1/4-acre or 10-acre lot. Because gardening and cooking go hand in hand, you'll want the healthy, fresh greens you consume every day to be close to your kitchen or back door so you can simply go outside and gather what you need.

There are other fruits and vegetables you will use, but not every day. Since, for example, you won't be collecting potatoes or squash every day, the staple area of your garden may be located further from the house but not too far out in the distance. The closer something is, the easier it is to control, maintain, and keep an eye on it. Additionally, your home, tool shed, workshops, and all other necessary garden infrastructure will be close by. In other words, easily reachable, saving you even more time and effort.

Animal/Pet Exclusion

Fences are the best way to keep deer, rabbits, and other animals away from plants. The size of the animals determines how big the fence should be. A three-foot chicken-wire fence can keep out smaller animals like rabbits. Fences must be at least 6 to 8 feet tall for deer.

DESIGNING YOUR GARDEN

Small-scale design layout

When it comes to your permaculture design, you are at liberty to construct whatever design you desire. I'm going to show you how the pros do it and everything they take into consideration.

Relative Location

The word "Relative Location" refers to the first permaculture design principle.

Every element (or individual component in a design) is positioned in relation to one another according to this principle so that they can support one another.

In permaculture, we place more emphasis on how things relate to one another and interact with one another than on the objects themselves.

Therefore, in permaculture design, we emphasize the inter-relationships between things, and by comprehending the nature of the elements and how they contribute to one another, we can choose the best position for them.

All elements have several inputs and outputs, which they are all capable of having. The outputs of one element in our design can flow into the inputs of another one if the elements are positioned correctly.

Understanding an element's nature will help us identify its inputs and outputs, and after we achieve that, we can decide where it should be relative to other elements in our design.

1. If you position water tanks uphill from your home, you won't need much energy to transport the water back down. This way, gravity does the work for you.

2. By planting deciduous trees on the sunny side of the house (the south-facing side in the Northern Hemisphere and the north-facing side in the Southern Hemisphere), we can block out the sun and thereby shade the house in summer when the tree is in full leaf, keeping the house cooler. Deciduous trees lose their leaves in autumn when the weather outside gets cooler. Since the leaves are gone, the sun may enter the house and warm it.

3. To make it simple to reach common vegetables and culinary herbs when cooking meals, a kitchen garden can be placed adjacent to a kitchen that opens to the backyard. In this instance, food is the kitchen's input from the garden. The kitchen scraps can easily be disposed of in a worm farm that is close to the kitchen garden. Kitchen scraps, which comprise the kitchen's output, are fed into the worm farm. Worm castings give the kitchen garden a nutrient-rich fertilizer that promotes the growth of the garden. Worm castings,

the worm farm's byproduct, are the input for the kitchen garden. There is no waste since, as you can see, all the inputs and outputs are connected across all three parts, and recycling is taking place as a result.

Kitchen Scraps Cycle

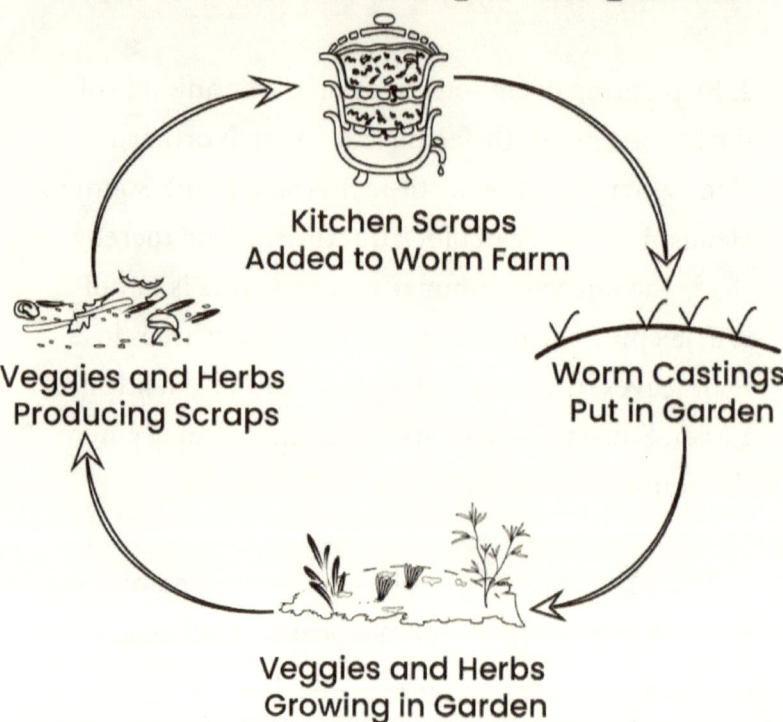

Kitchen Scraps
Added to Worm Farm

Worm Castings
Put in Garden

Veggies and Herbs
Producing Scraps

Veggies and Herbs
Growing in Garden

4. North-South oriented trellises are constructed to ensure that plants trained down its length do not shadow one another out and that the midday sun shines down the entire length of the trellis, maximizing exposure to the sun.

5. When utilizing the Relative Location principle, we consider the wind, sun, rain, and other environmental factors. We consider wind as an element in this example. Insect-repelling plants like wormwood can be planted upwind of our garden beds once the direction of the wind has been identified. The wormwood plants' aroma, which repels insects, is picked up by the wind as it passes through and mixes with the scent of the plants in the garden beds, some of which may include foods that pests could find tasty. Having been obscured by the stronger and more repulsive wormwood aroma, any bugs downwind won't be able to differentiate the scent of the veggies and won't follow the scent trail upwind back to the crops. This is one of the methods employed in companion planting known as "scent masking." In this situation, wormwood is preferable to other insect-repelling plants like tansy since it is an evergreen shrub that keeps its size and leaves all year long, whereas tansy withers and dies in the winter. Wormwood also provides protection for your winter vegetables. This is one of the things that makes permaculture so fun. It's exciting to know that you can use ways nature naturally behaves to do work for you, so you can spend less time toiling in your plant beds and more time enjoying their beauty!

6. By planting fruit trees alongside a chicken run, new food is made available to the chickens in the form of

fruit that has fallen from the trees, allowing us to recycle the waste. Fruit flies are not attracted since the fruit doesn't have time to decay. The soil beneath the chicken run is accessible to tree roots, and this soil is rich in nutrients that the tree roots will absorb. The hens can also find cool shade in the trees, and the chickens can control any insects that fall from the branches as a nuisance. Again, let nature do the work for you!

It's critical to keep in mind that "elements" in our designs don't just refer to the things we add; they also incorporate preexisting structures. These consist of structures such as trees and buildings as well as the "actual elements" of nature like the sun, wind, and rain, as well as diverse geological features like the soil type, slope/gradient, banks, gullies, waterways, hills, mountains, and so on.

In conclusion, we can use the Relative Location principle to optimize our designs by placing design elements close to one another such that their inputs and outputs flow into one another or where they interact with other elements to get the intended result.

5 WAYS TO INCORPORATE PERMACULTURE INTO YOUR GARDEN DESIGN

1. Build Healthy Soils

The best resource for a grower is soil. Organic materials and useful microorganisms are found in healthy soil. It effectively regulates nutrients and water, resists eroding, pests, and disease, and overall offers a happy environment for your preferred crops.

Consider the demands of the soil as a healthy, thriving basis for growing crops as your first priority with focusing on what a crop needs to be content as your second. Understanding the special conditions of your garden can help it thrive, as can being inventive and resourceful.

2. Make Wise Use of Water

An important component of a permaculture-designed garden is a strategy for effectively managing water. When water drains off your property or into the sewer, is this a waste of resources? Where it isn't wanted, is it creating erosion or pooling? These issues can be resolved by using a variety of water management techniques from the permaculture toolbox, such as water slowing or catching. I'll go into more detail in the next chapter on specifically how to use permaculture design to use water wisely, cut down on watering times, and establish the ideal moisture levels for healthy soil and crops.

3. Take Advantage of Plants' Many Uses

In the garden, a variety of plants can fulfill a variety of purposes. For instance, cilantro (also known as coriander), a famous edible herb and spice, also draws helpful insects and pollinators while producing an edible crop. Soybeans, however, may fix nitrogen in the soil while producing an edible harvest. By layering together multipurpose plants, you can make a more ecologically diversified garden.

4. Take Care of the Edges

The permaculture design process includes managing the edges of garden space. In order to effectively manage what enters the property, such as weeds, pests, wind, aerial chemicals, or water, the edges must be clearly defined. Perennial flower borders or hedgerows can beautifully frame the perimeter of a garden. They shield it from adverse weather and offer a habitat for helpful species, such as bees, toads, and snakes.

5. Guilds for Designing Plants

A plant guild is a collection of plants that help sustain each other. If you want to increase the harvest from a tree or provide support to it, think about growing extra plants beneath it. Examples include plants that produce mulch, serve as fertilizers, ward off pests, or draw advantageous insects. In order to increase production while decreasing labor and the need to import resources, a guild integrates plants.

DIVIDE THE GARDEN INTO ZONES

One strategy the pros apply for maximum ease and productivity is to segregate their space into zones. Zoning places the design's parts (such as herbs, trees, and chicken coops) in certain locations based on their function or our own needs. We position a piece closer to where it is needed or used most.

Getting Into the Permaculture Zone

Five separate zones make up permaculture (technically, six if you include zone 0). The exact limits of the zones are a moving target, but the general concept is that zone 1 is the place you visit every day, with each higher zone receiving decreasing amounts of traffic. Furthermore, elements in zone 1, such as crops, require routine maintenance, whereas, in zone 5, elements may not require any upkeep at all. Not everyone will have all of these zones where they are. Actually, the majority of people won't have zone 5. But that's totally fine. There are as many different ways to design a permaculture garden as there are people practicing permaculture. The important thing is that you make the layout that works best for you where you are now. I have seen just as many great sprawling rural farms as I have smaller permaculture urban gardens. Truly, they can be as amazing as you make them.

Permaculture Zones 0-5

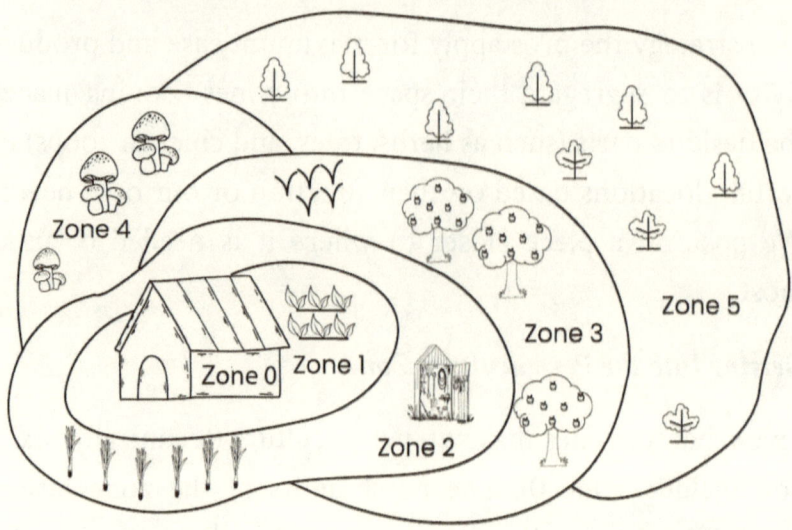

Permaculture Zone 0

While this zone is not always included in the traditional zones of permaculture, I think it is still worth mentioning. The house is in Zone 0. Food production (sprouts/ferments), trash management, maintenance, and education are all done indoors.

Permaculture Zone 1

Your front and rear yards are located in permaculture zone 1, which you frequent daily. You might walk through this location en route to your mailbox or your car. The secret is to figure out which regions you frequently visit or might visit with minimum effort. This space shouldn't be too large. You can only travel across so much space each day.

What then do you do in this area? Here is a list of zone 1 components that I like, but you can adjust these to what works best for you and your family:

- The Eco-Lawn (an environmentally-friendly lawn)
- Playground for kids
- A place to gather outside
- The kitchen garden (vegetables you often use in your cooking)
- An herb garden (herbs you often use in your cooking)
- Berries that are consumed frequently
- The lovely plants and flowers you love to see every day
- Wildlife-friendly native plants (I like these in every zone)

Although the components of your zone 1 may differ, I think you can see that this makes for a great space to have fun with family and friends, unwind, and gather food. Also, when you position your plants that need more attention where you are more frequently, you are more likely to spot issues before they go too far. Using permaculture zones in design makes your life easier, not harder.

Permaculture Zone 2

Zone 2 is the region of your property that is a little further away but that you still frequently visit. Instead of just visiting them without thinking, like with zone 1, you usually have to choose to go to these places.

Adding more fruit trees, berries, and perennial vegetables here may be an excellent idea. Additionally, beneficial additions to this permaculture zone are chickens and rabbits.

You don't have to be present all the time, even though you might visit these components every day.

Therefore, what components does permaculture zone 2 often contain? To help you get started, below is a list of some components.

- Small animals like rabbits and hens
- Vegetable plots that are devoted to crops you don't harvest every day, such as onions and a number of perennial vegetables
- A small food forest is a wonderful addition to zone 2, with dwarf fruit trees and berry bushes
- The worm farms
- Piles of leaf mold and compost bins
- Mulch mounds
- Plants that draw pollinators
- The main shed or store
- Native flora that benefits wildlife

Make sure zone 2 is simple to get to. This necessitates considering travel convenience in addition to distance. For instance, a steep hill may change the zone of a location near a house from zone 2 to zone 3 or even zone 4.

You won't spend much time there—at least not regularly—if getting there is difficult for you.

Permaculture Zone 3

This is the last regularly managed permaculture zone. Most likely, you won't be stopping by this location every day. In addition to nut trees and fruits like cherry, apples, and pears that may be collected in large quantities, this is an excellent location for your staple crops like wheat, corn, and potatoes.

In this area, animals who don't require daily care can also be maintained.

The following is a list of potential components for this permaculture zone.

- Food forest (mainly perennials; probably won't have typical garden vegetables, but perennial veggies, berries, and your larger fruit and nut trees are an excellent alternative)
- Large animals like cattle that are raised for meat
- The raising of mushrooms
- Beehives
- Food-producing crops like wheat and corn
- Native plants for wildlife

Despite being overseen, no one regularly visits and maintains this area. Make sure to avoid placing anything in permaculture zone 3 that requires your attention more frequently than a couple of times per week.

Permaculture Zone 4

Despite the fact that you can occasionally manage it, this permaculture zone is primarily wild. Zone 4 is where you can gather food from the wild and collect timber from larger trees.

With the help of birds and wind-blown seeds, nature is probably accomplishing most of the planting in this area. You can still choose which plants you want to allow to grow, though. This permaculture zone has more wild activity going on rather than you actively trying to produce abundance.

Here are a few things to do in this permaculture zone:

- Harvest large amounts of wood
- Scavenge for wild plants and mushrooms
- Hunt fish and other game
- Plant native plants for wildlife

Zone 4 is likely to feel somewhat untamed, but you'll be going there frequently enough to make an influence on a regular basis. Even so, this area offers crucial habitat for animals, which will help the rest of your land.

Permaculture Zone 5

Such wilderness areas are excellent places to view and absorb the natural world. This is the region that is intended to be entirely untamed and unaffected by human activities.

In actuality, this is one point where I disagree somewhat with the traditional permaculture zone breakdown.

Most of us won't be fortunate enough to live on a property that contains a truly natural region that humans haven't recently harmed.

You might ignore earlier human activity's detrimental effects if you simply leave this area. By taking a few simple steps, you might be able to encourage nature to restore a wild-looking degraded region into a fully functional and diverse wild area that supports more animals than it otherwise would.

The best course of action in this situation could be to initially classify a location as zone 4 and then, as you scale back your operations, move it to zone 5.

However, zone 5 of permaculture is also meant to be a place for research and observation. You might learn valuable lessons for the other zones by seeing how nature restores degraded wild areas.

Your choice of strategy is mostly determined by your personal ideals, although the following components and activities fall under zone 5.

- Nature observation
- Minor restoration of prior harm
- Very little harvesting (i.e., the occasional snack on berries and potentially taking plant cuttings for propagation)

Permaculture Zone 5 is, nevertheless, fundamentally a place where you go to watch and learn from nature. There won't be a zone 5 on every property, but that's ok. Work with what you have and you will still be able to develop a very satisfying and enjoyable permaculture garden.

Your garden shouldn't be a plain old piece of land, nor should it be an unsolvable labyrinth; the right design is the key to an easy permaculture journey. This chapter has shown you how to accomplish that and have your farm looking pleasant to the eyes and easy to harvest.

The design can make your garden look only so fresh; however, maintaining the aesthetics takes more than that. Your space would need a steady water source to keep up with its bright and healthy looks.

5

STEP 4: SET UP YOUR WATER MANAGEMENT SYSTEMS

S o you decided to read on; you must want your garden to look as vibrant as possible. In this chapter, I'll show you how to accomplish that with the right water system.

It wouldn't be any fun if there was one stoic way of delivering water to your plants; after all, permaculture is all about discovering new methods that work best for every party involved. We'll cover all known, practiced water management systems and how you can get them up and running on your grounds.

Let's get right into it!

WATER MANAGEMENT

There are numerous aspects of permaculture, but one of the most intriguing is how it handles water. According to permaculture experts, intelligent landscape design frequently makes it possible to go beyond water saving and recharge groundwater supplies. Geoff Lawton's "Greening the Desert" project in Jordan is the most well-known permaculture water-conservation demonstration site. Where no one thought it was feasible, Lawton's team was able to quickly grow food using a combination of mulching, contour swales, micro-irrigation, and meticulous planting.

Permaculture water management

What water resources are near your property, and what needs specifically should you address first?

1. Determine your water needs, and how you intend to utilize the water you have collected.

Can we accomplish anything without a clear understanding of our objectives and situation? Of course not!

To determine what size of storage you'll need to create and, more crucially, whether they'll be feasible given your terrain and financial situation, you'll need to be clear on the goals you have for your water system from the beginning.

Consider your intended use of the water first: will it be for domestic consumption, cattle, irrigation, fish production, fire protection, recreation, etc.?

After this, determine how much water you'll need for each activity by making an educated guess.

Consider what you can actually develop as the last step. Budget, available space, and aesthetics are all things you need to take into account in this situation.

This kind of thinking can help you prioritize based on the realities of your situation and can help you cut out a lot of needless planning. Because of this, we always establish your context and objectives clearly: Starting a project with a plan is the easiest method to avoid having to spend extra money on it later on!

2. Recognize the water sources.

Let's look at the water sources your farm has access to after you have a better sense of your water requirements and how you want to use the water you harvest. As I mentioned in the last piece, you can learn this by doing an internet search and observing your surroundings.

Therefore, the first thing you want to ask yourself is: How much precipitation do I get annually? More specifically, what is my typical annual rainfall, expressed in millimeters or inches? Second, how is that rainfall dispersed throughout the course of the year? Is it being supplied in torrential downpours, only in the winter, or equally dispersed all year?

Based on these figures, your water systems will differ greatly. It will take a completely different strategy if you get 50 inches (1270 mm) evenly spread throughout the year versus if your average rainfall is 25 inches (635 mm), and most of it occurs in a few intense summer storms.

The basis of your strategy will be the precipitation and its distribution, and you can quickly acquire this vital information online with only a few mouse clicks. You'll now need to conduct some permaculture sleuthing to determine potential water sources on your land and beyond.

Any streams that traverse your land should be noted and distinguished. In essence, this flowing water represents runoff from areas of your watershed that are outside the borders of your land. Although you have no control over how this water enters your property, you may use it to meet your water demands if necessary. You must understand the exact dependability of your water source for this reason. Is it perennial or only intermittent? When there is a drought, can you rely on it?

Finally, take into account whether you have access to underground water. If you don't already have a well, you can't confidently estimate how much water you'll have under your feet unless you drill one. A local company that specializes in wells can assist you with this. If you have groundwater, you should consider using that as a supply of water.

3. Watershed - identifying your position within the hydrologic cycle and the watershed of your site.

Now that you are aware of the precipitation you receive and the additional water sources you have access to, you can begin by examining your watershed and figuring out where you fit into the hydrologic cycle (also known as the water cycle).

A watershed is a region of land that drains rain or snow runoff downhill from the highest topographical barriers, such as hills, ridges, and mountains, to a particularly low point, usually a tributary outlet to a bigger river or a lake.

Your land almost certainly forms a small portion of a bigger regional watershed that drains hundreds of square miles or kilometers of land to form streams and rivers. I would advise that you initially consider the larger watershed even though understanding your local watershed might not be immediately useful to you.

Your location within the larger watershed affects how much water moves on your property or in your neighborhood. For instance, if you're up in the hills, there will likely be a few small creeks, but if you're low in the landscape, there will likely be a lot of water—probably rivers rather than creeks—flowing through the area.

However, you must consider your site's watershed or sub-watershed if you want to utilize its actual water resources. Although you may be a part of a sizable watershed, the exact amounts will depend on the area's topography.

Understanding the terrain patterns shown on topographical maps is essential to this process of determining your site's watershed. The easiest way to do this is to do an internet search for "topographical map", select the top option, and type in your location. When I did this for my location, I thought it was fascinating to find so much information on the area right around me that I had no idea about before. Looking at this topographical map will help you identify your land's contours for their definition of ridges, saddles, and valleys/gullies.

WATER STORAGE

There are a number of ways you can have water stored for your farm; we'll look at them all, from the simplest to the most complex.

Let's begin by storing the water in the soil, shall we? After all, the earth is the cheapest place to store water.

Storing on the Soil

Let's first take care of the necessities that won't cost too much money, even though you may have grand dreams for an interconnected network of cascading ponds. The largest storage resource present on the majority of sites is the soil. Getting water into the earth and storing it there should always be our top priority.

You must concentrate on two goals if you want to keep water in the soil. The first is to slow down, disperse, and sink the rain so that it travels the furthest distance possible across your property, running to as many objects as possible, dispersing where it's needed, and allowing time for it to permeate until it inevitably leaves and drains away.

The soil's organic matter is crucial to its ability to hold water; therefore, your second goal is to increase its amount. The organic stuff progressively absorbs the water in the landscape by acting as a sponge. Therefore, encouraging topsoil rich in organic matter is essential if you want the soil to hold more water.

The difference between poor soils with less than 1% organic matter and soils with even as low as 2% organic matter is staggering. Just that much change can reduce the demand for irrigation by 75%. Suffice it to say, soil sponge development should be a top priority.

Additionally, you must design the terrain so that water will spread slowly and sink, allowing it to be absorbed by the sponge. Two extremely well-known methods can be used for this: Keyline subsoiling and Swales along the contour are two examples. Let's begin with the keyline.

Keyline

Thanks to P.A. Yeoman, the idea of keyline agriculture was developed in the drylands of Australia. This now-famous Australian man has influenced the way permaculturists think about managing water on the farm.

While keyline agriculture encompasses a variety of ideas, its most basic principle is the distribution of water from locations where it is concentrated in wet areas to areas where it is constantly too dry. See, water typically flows from slopes into valleys. While the valleys pick up moisture, the hills remain dry.

But if you use a keyline cultivation scheme, you may direct the water away from the valleys and toward the ridges, distributing it evenly throughout the land and enhancing infiltration. This is accomplished by utilizing the tractor to rip lines (open up soil furrows) while employing a keyline plow that is parallel to the keyline (thus giving the name keyline cultivation pattern).

The water will then be diverted from flowing down toward the lowlands and moved in the opposite direction, toward the ridges, by these tiny water channels and drains that have formed in the soil. Overall, this causes rip lines to retain water for infiltration rather than allowing it to flow down the slope. Plant growth and soil microorganisms both rise in response to soil moisture.

Keyline Cultivation

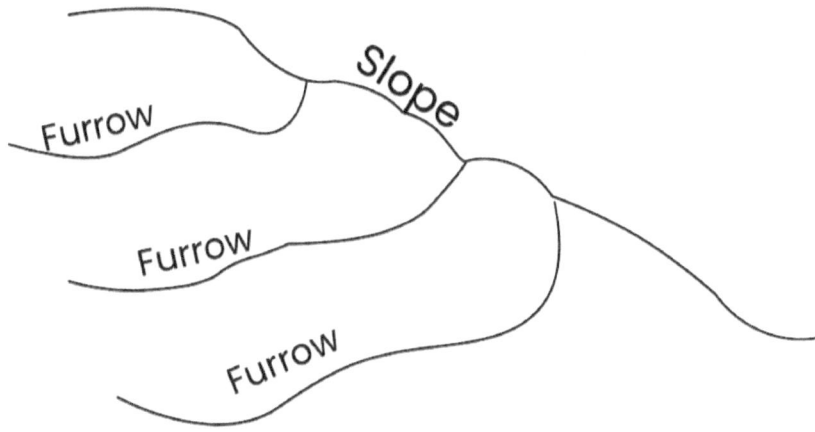

Given that keyline cultivation encourages the quick production of topsoil, it is also a strategy for improving the soil. By making furrows in the soil and ripping up the subsoil, you can get water and air deeper into the soil, where plants can use them. This can soften the hard pan and create rich, fertile soils. As you are already aware, fertile soil can hold and absorb more water.

Therefore, it is clear why keyline pattern cropping is a fantastic tool for controlling water on a permaculture farm. It may collect rainwater, evenly distribute it, and create lush, rich soils by turning subsurface into topsoil.

Swales Contour

Swales are your second method for keeping water in the soil. Swales also aid in slowing, dispersing, and sinking water, which enables us to hold back runoff and allow it to seep into the soil, where it is then stored.

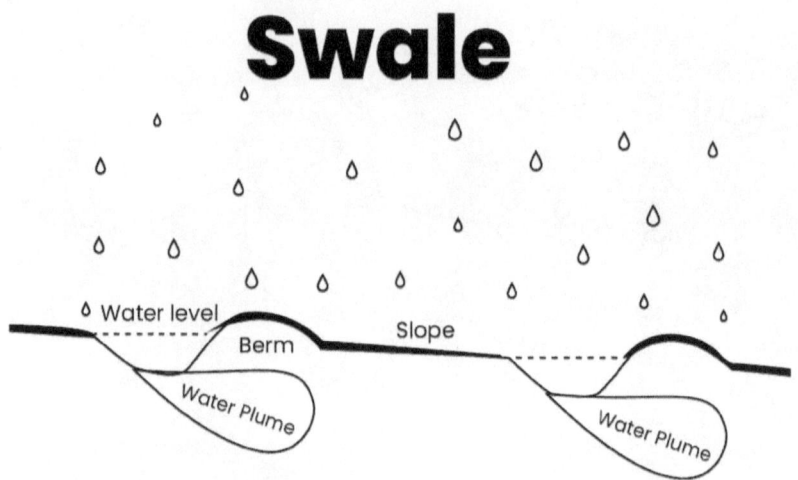

Toby Hemenway defines a swale in his book *Gaia's Garden* as "a shallow trench laid out dead level along the land's contours." It can range in size from one to several feet long, a foot or two deep, and any width required. A raised mound or berm is created by piling the earth excavated from the swale on the side facing the downward slope.

When the earth can no longer absorb the rain during the rain event, overland flow develops. Runoff is the downward movement of any water that the soil is unable to absorb. Surface water and rains that are flowing downhill are caught by the swale, spread out throughout its length, and gradually soak into the soil.

Following that, a lens of moisture is created as the subsurface water seeps downslope. Tens of feet below the swale, the water that is stored produces an underground reservoir that promotes plant development. Swales are, above all, tree-growing systems. When you plant trees or other crops on the berm (mound) on the downhill side of the swale (or just below it), they will be able to benefit from this soil moisture during dry periods.

Swales are the primary tool we employ for this, but they also help to prevent gullies from forming by catching rainfall, slowing it down, spreading it out, and essentially lowering its erosive potential. Swales also collect organic debris, transforming the ditch's surface into a thick, rich layer of humus that may hold a sizable amount of water. Additionally, once it has been dug out, you can fill it with wood chips and dead branches to add organic matter.

I know you'll be anxious to use swales on your land after reading about them, but will they be effective there? Swales are the most popular and misunderstood permaculture water management method. Because some regions are more likely to suffer powerful storms with greater runoff, the climate of the area, soil, and slope all affect the situation. You don't want to place a swale on a steep slope and cause a mudslide. Also, keep in mind where you place your swale in reference to other structures on your land. It's a great thing for your trees and other vegetation to be right above a swale's water plume. However, it is not a great thing for your house's foundation.

Storing Water on the Surface

Okay, now that you've finished creating that inexpensive subsurface water storage, let's move on to storing water above the surface. Here, we'll create water systems for storing, collecting, and recirculating surface water.

Storage of Water

Water Tank

A water-storage tank might be a more cost-effective and advantageous choice if you only need to store drinking water. A tank is a preferable choice once more when your site's topography may suggest that building a pond would be too pricey.

You can build water tanks out of a variety of materials, and if you place them somewhere on the top of your space, at the highest practical point, you'll be able to use gravity to bring the water down to you when you want it rather than using energy to pump it to you.

Pond or Dam

A dam or pond used for water storage is the most affordable place to keep huge amounts of water. A pond's water can be used for various things at once, including aquaculture, irrigation, stock, household storage, wildlife habitat, recreation, and more, making it a very valuable resource in a changing climate.

The two main types of ponds/dams are an embankment pond and an excavated pond. An embankment is created by placing a dam or embankment over a stream or waterway where the stream valley is sufficiently depressed to allow for the storage of enough amounts of water, as the name implies.

A trench or dugout must be dug in a nearly level area to create an excavated pond. Excavated ponds are utilized when a limited amount of water is required because the water capacity is virtually totally achieved by digging. Some ponds are constructed in gently to moderately sloping regions, and the capacity is achieved by combined excavation and dam construction.

The type of pond you'll be able to build on your property will depend on the topography of your land. The most crucial element in assessing a possible site's success is the storage ratio, which varies depending on the type of pond and the location (the volume of excavation versus the volume of storage). When building a pond, you want to make sure that you devote the least amount of labor and earthwork possible for the most quantity of storage.

The pond's size and type will also be influenced by the climate and the average evaporation losses. In comparison to milder climes, evaporation will be fairly high in semi-arid and arid zones. In order to compensate for annual evaporation losses, ponds in the hotter zones must be deep.

In light of this, let's go over the various pond varieties, starting with the cheapest and easiest to dig up to the priciest ones that demand more involved earthworks. When doing so, the first guideline of working with water is to keep it in the location on the landscape where it has the most potential, preferably high up if it can be done so inexpensively. Therefore, we'll begin at those points and move downhill from there.

Gully/Keypoint Ponds

These are perhaps the most prevalent dams and one of the simplest methods for storing water. They are also the most affordable choice because they are built by creating an embankment in a drainage depression or gully. Building a dam wall that can contain water in a valley or gully behind it is the crux of the necessary earthworks.

Keyline design concepts would once again be the correct approach to determine the ideal pond position in the gully/valley. This entails first locating the main keypoint of the slope (when the gully/valley slope section transitions from a concave to a convex profile). Once the keypoint has been discovered, the keyline is the contour line on the terrain that passes through the keypoint.

This keyline is said to be the highest contour in the gully or valley that can hold water effectively. It is typically the highest point in the landscape where it is practical to hold water. Water for irrigation is mostly stored in keypoint dams and ponds. This irrigation water is typically released through the sizable pipe beneath the dam's wall.

Saddle Pond

A saddle is a type of topographic feature that is nothing more than a dip or break along a straight ridge crest. This area has the highest accessible water storage because it is on a ridge and is therefore on higher ground. Compared to a gully or valley pond, this one has a considerably smaller watershed, yet it can still catch runoff from both ridge crests. A saddle dam's main purposes are for domestic livestock and wildlife, not irrigation.

Hillside/Contour Ponds

Built on the sides of hills, contour or hillside ponds typically have a long, curving bank that runs straight across the slope of the hillside or a three-sided or curved bank (on the contour). Search on your topographic map and look for any expansion of the contours along the hillside to find these types of dams. Widening indicates that the land is flattening, making this a potential site for the pond.

These ponds are more expensive to construct because more digging is required for less water storage, but they will still give you gravity storage. Ponds in the flat are still subordinate to gravity-fed water. They are used for domestic cattle and wildlife and are often filled by graded catch drains, or diversion drains.

Ponds for Flat Sites

Excavated tanks, ring tanks, and "Turkey's nest" ponds work well for flat areas, but since they can't collect runoff, they must be filled from outside sources. In tanks that have been excavated, the excavation is now used to store water below the surface. Unless new dam walls are built to provide extra above-ground storage, the removed earth is heaped nearby.

Ring tanks are built by constructing the surrounding embankment out of dirt taken from inside the ring, which can be circular or curved to fit the topography. Usually, water is kept above the surface of the earth. A "Turkey's nest" dam is a ring tank variation in which the borrow pit is outside the embankment. The storage of water is also above ground.

WATER HARVESTING

So you have a storage system in place; what next? You must then refine and improve your water collection techniques. You can use a well to fill your ponds, but you should first use surface flows and rainfall runoff to fill your water storage before you dig down and access the subterranean aquifer.

With the use of water-harvesting drains, you can collect water by directing runoff, stream flow, or pumping water into your ponds and water tanks. Bill Mollison says in his *Designers' Manual*, "The purpose of these drains, which are actually ditches in the soil with a modest gradient, is to direct water to a particular area, such as your pond. They are installed off-contour in the landscape."

Consider diversion drains/ditches as enormous earthen gutters spread throughout the landscape that collects and transports water in a way akin to a home's rain gutters. Diversion drains operate best when the base and sides are clay-lined, unlike swales, which are typically built on permeable soils. They differ from swales in that they are designed to flow after rain.

On the other hand, swales or ditches on contour can also be used to collect water for you, and if they are connected to a pond, they will overflow into the ponds as they fill up. Additionally, if you have several connected ponds by swales, the overflow from one pond will reach the feeder channel or swale of the next. A spillway is necessary for a pond since you may then slow, disperse, and sink water across your area.

Your roads themselves become a very significant and effective water-harvesting system once they are installed. The roads have a very high runoff coefficient due to their compacted, graded, and frequently impervious construction. The road will be the sole surface for runoff in some environments. When combined with other harvesting drains and/or swales, the roadways and nearby water collection drains can help hydrate the farm as a whole.

Water Distribution

Stored water must get to the target destination somehow, and it would take a very long time if you had to use buckets and watering cans. This is why you must consider a standard or improvised water distribution system.

There are two main water channels to consider while constructing a farm's water resources, in addition to the previously stated harvesting drains. For irrigation, there is another type of water channel. These harvesting drains, also known as diversion drains (positioned in the landscape off-contour, with a little gradient transporting the water), serve in this instance as irrigation channels for flow irrigation.

When a pond's water is poured into a drain, it fills up and overflows over its whole length before cascading over fields of crops or paddocks. In essence, you may use these kinds of drains to water your pastures or irrigate fields with crops like potatoes, corn, and beans.

Shawn Jadrnicek suggests in his book, *The Bio-Integrated Farm,* that in order to construct this drain, one should start at the outlet, which could be a pond, retention basin, swale, or some other area with the ability to hold and safely release the harvested water and work its way down the slope towards the desired irrigation area.

A gravity-fed pipe network can release the water that has been kept in ponds and water tanks as another method of moving water. To use this strategy, you will use your header water tank, situated at the highest point in your landscaping, and release the stored water to irrigate your gardens and orchards using the irrigation system's pipe network. By using a pond as a reservoir and a network of irrigation reticulation pipes connected to it, you might accomplish the same thing on a larger scale.

Apply Water Properly

You want to water your plants enough, but not too much. When you're starting out, it can be tough to find that perfect balance, so I'll give you some advice. In general, you don't want the ground to dry out completely. If the soil is looking dry and dusty, it really needs a drink. Another tell-tale sign that a plant needs more water is if it starts to droop, wrinkle, or turn brown on the edges. If you catch it soon enough, a good watering will revive these plants. However, if you don't water them and you don't get a good rain, before too long those plants will die. But don't take this to mean that more water is always better. Plants can also suffer from getting too much water. If the soil is constantly overwatered and soggy, you'll find yellowing leaves, root rot, wilting, and leaves turning brown. Yes, some of the symptoms of underwatering and overwatering are the same. So, if you see those signs in a plant, check the condition of the soil to determine the problem. Plants also need their leaves to be able to dry between watering, or they can develop disease. If you're having a problem with the ground constantly drying up while the foliage is still damp and doesn't get a chance to dry off, you have a couple of options. You can prune or you can move the plant a little further away from others around it. Either way, the goal is to allow more room for wind to blow around the plant and dry it off.

Another thing to consider is that different plants like different conditions. While what I said is generally true, there are exceptions, such as desert plants that do very well in dry conditions. So, if you're having problems, check out how much water that specific plant prefers.

With this guide, you should be able to decide exactly which water system best suits your farm. You have to consider the system, the size of your farm, and how many resources you want to pour into it. But once all those are sorted out, you should have your water system up and running in no time!

After all that work, wouldn't it be nice to just find a great-looking bed where you can rest? I will be talking about beds in the next chapter, although it is not exactly the kind you want to get comfortable in.

6

STEP 5: BUILD YOUR BEDS

"A long, narrow urban garden can be difficult from a design standpoint, no matter where you live or what temperature zone you are in. Despite being only 21 feet [6 meteres] wide, this unique garden is about 100 feet [30 meters] long from north to south. The soil at the site is stony, loamy, clayey, and rich in lime, and its drainage is only marginally hampered.

Winter lows hover around 34 F [1 C], and summer highs average around 70 F [21 C]. Approximately 24 inches [610 mm] of rainfall each year, and while water shortages are often not a big problem, they are becoming more frequent in the spring and early summer.

But the main reason the client came to me for the design was to advise her on a layout and design that would enable permaculture in practice and provide a space that the entire family could enjoy, *as they had not been using the garden much up until that point, especially the end that was farthest from the house.*

The ideal arrangement of the design's various components was chosen using permaculture zoning. I advised building the first garden room, the kitchen garden, in zone one, just beyond a patio, outdoor kitchen, rainwater collection system, and composting area. This area's herb and flower edging served as a space divider.

I suggested building a little wildflower meadow adjacent to the kitchen garden, complete with washing lines for drying clothes. A tiny greenhouse or polytunnel for year-round growth is located immediately beyond this. This structure also breaks the sightline, which helps the garden appear less long and narrow.

Around half of the area is taken up by Zone 2, a lush forest garden with a trail leading to a wildlife pond and a pergola-covered patio next to a summer house.

Mixed hedgerows along the east and west edges are considered zone two and offer a variety of yields, both edible and not.

Lastly, a tiny, natural space behind the summerhouse at the far end of the garden, behind mature trees, is to be mostly unaltered for animals. Nevertheless, it might also make it possible to grow mushrooms.

This design uses permaculture zoning to create a practical garden where the areas that are used the most frequently are located closest to the house. However, it also promotes the use of the entire garden by positioning the summer home as a "destination" at the end of a succession of lovely garden rooms."

-The Story of A Long, Thin Garden in England

The next thing you need to work on in your garden is the beds. I'd go on about how vital this stage is, but the story already did that justice, I'm sure.

But I'm going to buttress the point anyway; after all, you paid for convincing!

BENEFITS OF A RAISED BED GARDEN

Every year, the popularity of raised bed gardening rises. Raised beds come in a variety of designs, sizes, and price ranges, ranging from straightforward, cost-effective structures to beautiful, intricate pieces of garden art.

Why are raised bed gardens so enticing when farmers have long-grown vegetables in the ground? Let's find out!

Super Soil

Raised beds firstly give you control over the soil. Raised beds offer ideal soil conditions, regardless of whether you struggle with clay soil or have had a disease in your garden that was brought on by the soil. You may manage the soil's composition and structure within the confines of the bed, providing your crops with a nutrient-rich environment. To keep the soil healthy and productive, it is practical to test the soil each season and add any necessary nutrients or compost.

Healthier Harvests

With numerous raised beds set aside for your kitchen garden, crop rotation may be easily planned and executed. Crop rotation promotes soil health and prevents pests from overwintering in the soil. In the spring, when newly hatched bugs emerge and discover that their food supply has shifted to another bed, they will make an effort to move to the new bed. Thankfully, most pests will perish on the trip, falling prey to birds or other predators.

Water Control

Raised beds prevent soil from becoming waterlogged, which is a problem for many in-ground gardening. Irrigation is simpler and less wasteful in the small area of a raised bed. This ensures healthy plants and reduces water usage costs.

Season Extension

You can plant your spring crops early because raised bed soil often warms up earlier in the spring than ground soil does. In addition, you can build low tunnels over your raised beds with the addition of straightforward support, extending your growing season into late fall or winter, dependent on your hardiness zone. You may even build a cold frame for winter gardening using your raised bed as a base by adding recycled windows to it.

Pest Protection

Raised beds can prevent crops from being used as animal treats. Voles and gophers cannot access your excellent root vegetables if the bottom of the bed is reinforced with a barrier, and cabbage worms can be avoided by using a row cover.

Prolific Produce

In order to increase the amount of planting space in a raised bed garden, vertical supports can be added. You can put lettuce beneath trellis-grown peas and grow radishes around the perimeter of the plot as a border. You can think about Mel Bartholomew's idea of square foot gardening, which employs a grid to split the raised bed into square foot increments to increase the yield of your garden. Then he illustrates how many plants can be grown in each square, such as one tomato per square foot, 16 onions per square, two pickles, and four ears of corn. To maximize planting and harvest in your space without overcrowding, the grid rests flat over the raised bed, eliminating the guesswork as to how many plants you can fit into each bed.

Aesthetic Desires

Raised beds give the landscape a visually appealing structure. Numerous homeowners want to cultivate their own food, but annoying Homeowners' Associations dislike messy gardens. Raised beds can be made aesthetically so that the crops can be contained, edible flowers and pollinator plants can provide visual interest, and the raised bed can become a lovely garden focal point. A gorgeous edible garden fit for a magazine photo shoot can be created by building multiple beds, placing them at regular spacing or in a potager-inspired design, adding stone pathways, and adding an arbor.

Raised beds have various uses in the garden, from easing some back pain to regulating the soil's composition.

Types

What materials do you need to build a bed? You actually have a lot of options in this respect. As long as you have your lumber and a bit of proficiency with drills, you can quickly and easily build a raised bed when equipped with a measuring tape, drill, and level. You can also use:

- Salvaged wood
- Cinder blocks
- Bricks
- Fallen logs
- Straw bales

THE OPTIMAL BED FOR GROWING VEGETABLES

Using the least disruptive way is one of the primary strategies of Bill Mollison's permaculture principles and Masanobu Fukuoka's natural farming philosophy when breaking ground and constructing new bed systems. You will save a lot of energy, encourage growth, and enhance your productivity by being aware of the many methods for making garden beds, including their pros and cons. It will lessen maintenance requirements, improve soil fertility over time, and generate a habitat.

There are several different garden bed concepts and terminologies. Some of the most popular techniques for growing food include no-dig and no-till gardens (also referred to as lasagna or sheet mulched beds), as well as double-digging garden beds, hugelkultur, raised beds, and spot planting. They are all related by referencing the idea of enhancing soil fertility while employing the least harmful technique. To create an overview that is simple to understand, I have organized the five various garden beds into approaches and shapes.

No-till, no-dig, lasagna, sheet mulched garden beds

When To Use

This style is known by many names and is ideal for keeping the soil biology that has already developed intact. Digging is not required, so microorganisms and soil food webs that have developed will not be destroyed.

How to Build

Lay newspaper a couple of layers thick on the ground and get it damp. Water is necessary for the decomposition process, and getting it wet also helps it stay in place while you're working. The newspaper will serve as a nutritious layer of carbon and will smother the grass, weeds, and anything else beneath it to make room for your new plants. From there, alternate layers of green material (grass clippings, leftover vegetables from the kitchen or garden, or coffee grounds) with layers of brown material (newspaper, dry fallen leaves, compost, or worm castings). The greens are rich in nitrogen, while browns are rich in carbon; both are crucial to the soil's health. As a general rule of thumb, you want about twice as much brown as green.

Make the bed taller than you want your finished product to be because the layers will get thinner over time as they decay. When you're initially building the bed and have reached a height of at least 6 inches (150 mm), cover with mulch, add water, and let those little microbes work on your behalf all fall and winter to make an excellent bed for you for spring.

Advantages

- The work is carried out by the natural process, which means the existing soil structure won't be damaged in any way.
- It requires the least amount of labor and uses supplies that you probably already have, such as grass clippings, kitchen scraps, and cardboard.

No-Dig/ Lasagna

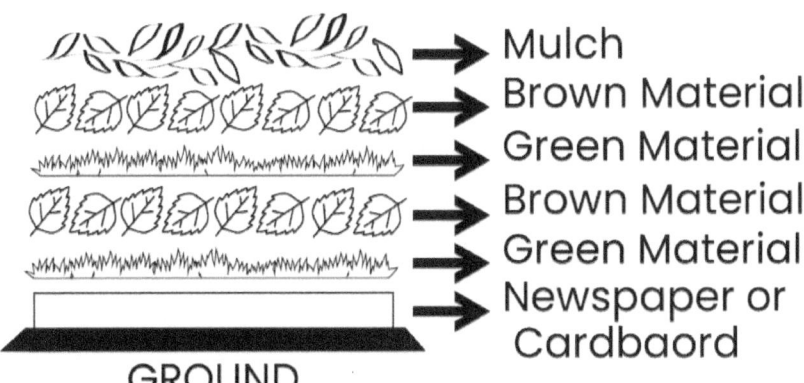

Mulch
Brown Material
Green Material
Brown Material
Green Material
Newspaper or Cardbaord

GROUND

Disadvantages

- If you want to plant right away, this type of bed is not the ideal option because you need to wait for the layers to break down. However, if you want to work around that, add topsoil to the mixture before laying mulch, and then you can plant right away.

Double-dig garden beds

When to Use

Although labor-intensive, this method is helpful if your soil has poor water drainage or retention and is filled with rocks or clay.

How to Build

The topsoil is removed first, then placed aside. Next, till the ground underneath while removing large rocks as you go. Leaving behind some small rocks or pebbles is fine because they don't significantly impede root development and are excellent moisture reservoirs. After that, incorporate organic materials like compost or vegetable scraps. Reapply the top soil, and then cover with mulch. Add more compost before mulching if you want to improve your top soil as well.

Advantage

- A quicker method of boosting soil fertility compared to the others.

Disadvantages

- A lot of labor is required for this method.
- This approach destroys any existing, sound soil structure. However, if the soil is especially poor, there won't be much beneficial soil structure. So, in that case, it would be worth it to dig and till.

Double-Dig

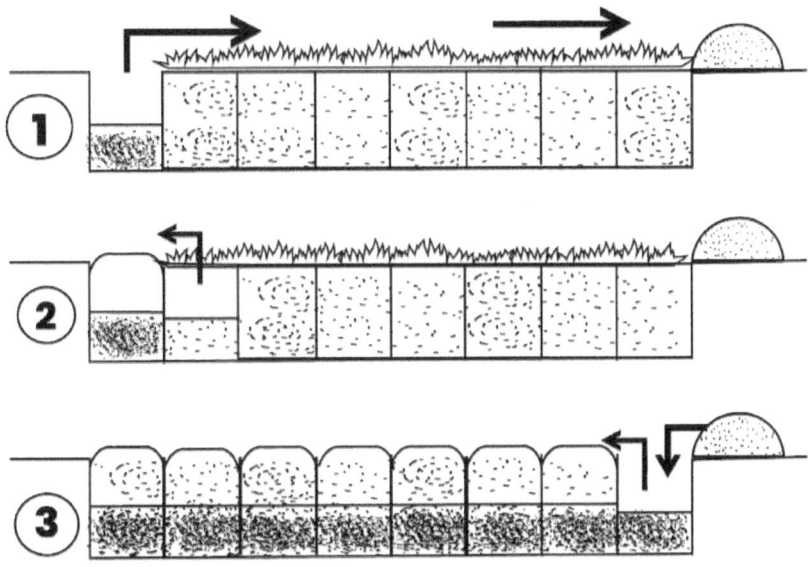

Raised and waist high garden beds

When to Use

Urban locations benefit greatly from raised beds because they may be built on concrete or virtually any other surface not ideal for gardening. Additionally, they support remediation efforts in areas with poor drainage.

Waist High Bed

Raised beds are ideal for inclusive gardening since they provide access for those in wheelchairs or who suffer from back problems.

Raised Bed

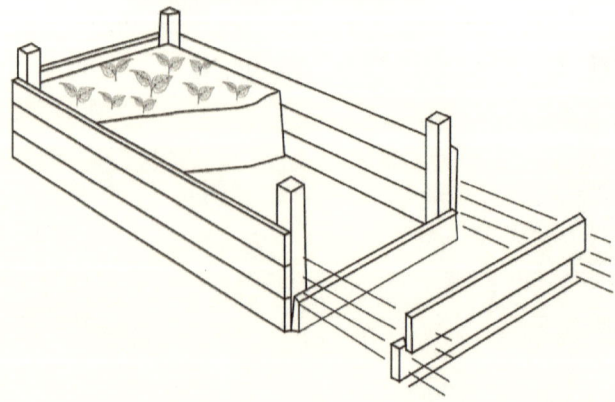

How to Build

A waist-high garden bed is basically a no-dig garden on stilts. The raised bed version is the same, but directly on top of the ground. After you build your frame, you can fill your beds with several layers of the brown and green materials mentioned earlier. If you want to use these beds right away, do only a couple thin layers of brown and green material and fill up your container the rest of the way with potting soil, then cover the whole thing with mulch. Onions, tomatoes, leafy greens, potatoes, and root crops generally do well on raised garden beds.

Advantages

- Free draining and useful in both urban and commercial settings.
- Long growth season and high fertility.

Disadvantages

- More expensive and requires more planning than the no-dig method.

Hugelkultur bed

When to Use

If you like to plan a few seasons ahead and want something that is going to fertilize itself for years to come, this is the garden bed for you.

This style of bed is helpful for very poor soil, such as inadequate nutrients or too much sand or clay leading in water retention or drainage issues.

Hugelkultur

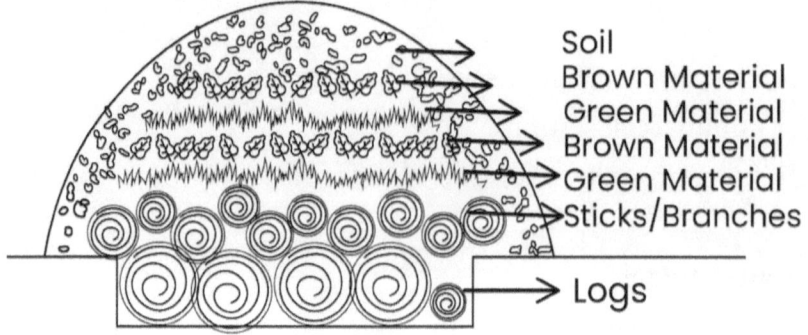

Soil
Brown Material
Green Material
Brown Material
Green Material
Sticks/Branches

Logs

How to Build

The German word "hugelkultur" means "mound bed" or "mound culture." The top soil is first dug up a few inches deep and placed aside. The truly remarkable part comes next. With logs or branches, you plug the hole. They will serve as a productive reserve and gradually boost soil fertility as they decompose. With tiny twigs or other brown materials, fill in the spaces. Finish the process as you would for a lasagna garden and then you have your "mound bed", or hugelkultur.

Advantages

- Long-term preservation of soil fertility.
- Produces a microclimate.
- Aids in improving drainage and moisture retention on the property.

Disadvantages

- High energy input and initial destruction of soil life and structure.
- Takes up more space.
- If you want to utilize the bed right away, you'll need to cover the wood with enough nutritious soil for the roots have room to spread downward.

Spot planting

When to Use

Spot planting is an excellent place to start when converting a large area of ground into a productive garden bed. The majority of the time, it begins with the creation of "islands"— fertile soil pockets for trees surrounded by vegetation that supports the tree's growth while producing a harvest (also known as companion planting). In order to build a growing oasis, you can then begin connecting the various "islands" from there.

How to Build

The same layering technique is used for spot planting as it is for no-dig beds. You are only paying attention to the area where you will plant a tree or make a little area fertile enough for the growth of vegetables. You can turn a little portion of your land into a garden by spot planting.

Advantages

- It's a good place to start if you're feeling overwhelmed by too many options.
- Gentle conversion of small plots of ground into productive and useable garden beds requires less energy and resources.
- Focused and minimal human interaction with existing nature.

Spot Planting

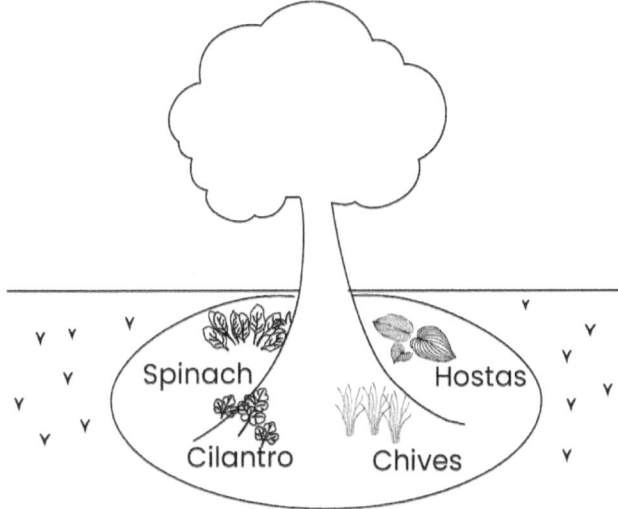

Disadvantages

- This is a very small garden, so it won't be able to supply all the herbs or vegetables you may want in your kitchen garden.
- The garden bed is flanked by wilder regions, making weed management a challenge.

Now you know the benefits of creating great garden beds, your options, and how to set them up. That's great progress! Once your beds are set, the next step is to start laying seeds and planting crops. If you're ready for that step, then follow me to the chapter!

7

STEP 6: PLANT YOUR GARDEN

In all the beautiful chaos of your own rebel farm, there simply must be a bit of order for it to thrive. When you want to plant your crops or seeds, it's not advisable to just throw seeds around, hoping they land on fertile ground; you made beds for a reason, after all.

In this chapter, I'll walk you through the best planting strategies which you can apply to your vision for your garden.

PLANT PERENNIALS FIRST

I had start with this point; that's how vital it is. The location of the perennials should be considered at the first planting stage of your garden because these are the plants that will return year after year and become a dependable part of the garden.

You can grow perennials from seed, but many of them, like asparagus or fruit trees, will take a while to bear fruit. Therefore, purchasing transplants or bare-root saplings is preferable to shorten the amount of time before you can begin harvesting them.

Let's look at some other reasons why you should start your garden out with perennial plants.

Low-Maintenance Perennials Means More Time For You

Perennial cultivars need far less maintenance than annual ones, whether you're planting a formal garden or a field of wildflowers. You can anticipate spending less time and effort feeding, watering, and caring for your plants overall. Fall and spring are both good times to grow perennials. Don't worry if you don't see blooms the first year if you water and frequently weed in the first season; perennial plants use this time to build robust root systems. Only during their initial growing season can perennial wildflowers and the majority of perennial plants display green growth. You'll have access to brilliant blossoms with minimal upkeep in their second and subsequent years!

Established perennial meadows and gardens need very little worry and care, although you should still weed them and supplement them with water when it's particularly dry. With less frequent maintenance required, you have more time to enjoy your landscape!

Perennials Give You More Bang For Your Buck

Perennial plants, as opposed to annuals, are a one-time investment, making them a wise financial decision for gardeners of all income levels. Many perennials, including daisies, lupines, daylilies, and others, grow more each year. You can divide naturalized perennials and replant the divided plants in different areas of your garden.

Look for perennial plants with long bloom periods or reblooming varieties for the greatest number of blooms!

Perennials Make A Big Statement

Perennial gardens and meadows provide the most powerful seasonal statement. You can create and organize your perennial garden such that it changes with the seasons. Strong hues can be found in flowers and foliage in a well-planned perennial garden from spring till October.

Essential steps to planting perennials

So you've decided to plant your first perennial, that's great! How do you do it, though? Let's find out.

Condition the Soil

Add as much organic matter as you can to your garden. Utilize materials like compost, old leaves, mushroom compost, bark mulch or fines, or composted manure. Adding a lot of organic materials establishes a basis that promotes plant growth. It's best to complete this task a few days, months, or even an entire season in advance. For spring planting, many gardeners prepare the soil in the fall. If you can stand the heat, prepare the soil for planting in the fall this summer.

Test The Design

Plants should be arranged in the beds before you plant them if you're planting a garden. Try to complete this a few days prior to planting so you can make any necessary design adjustments. This allows you to view the garden's design at various times of day and from various perspectives, both inside and outside. While pots are outside in the bed, keep plants well-watered. This also gives you the perfect opportunity to harden off the plants if they have been kept indoors and aren't used to being outside all day. For any plants that need this special attention, put them outside for just a couple of hours the first day. Every day going forward, increase their outdoor time a couple of hours at a time. Each time you bring them back into the house, check to make sure they still look happy. If they are exhibiting signs of stress, decrease their outdoor time until they are doing better, then start the process of increasing outdoor time again until they can stand being out all day.

Make a Planting Hole

Make the planting hole twice as broad and just a little bit deeper than the pot the plant is in. Give the hole a few handfuls of organic material. Water the hole before planting if the soil in the bed is dry. To avoid the soil in the hole drying out once it is exposed, drill planting holes in a large bed one at a time.

Tease Roots

Take the plant out of the pot. Tease and break up the root ball's base if the roots are thick and tightly molded to the curvature of the pot. New roots are induced as a result. Put soil and teased roots in the planting hole to add organic matter.

Check the Depth

Place your plant in the hole. Verify that it is facing in the desired direction. Examine the planting depth. Try to place the plant in the hole at the same depth that it was in the pot. Do not bury the crown, the intersection of the roots and stems. Laying a stick or other object across the planting hole from the surrounding soil to the perennial root ball will allow you to quickly gauge the depth of the planting.

Fill the Hole

The soil you dug up from the pit should be mixed with organic material, such as compost. If desired, add a small amount of granular organic fertilizer. To start filling the hole, combine everything and toss it like a salad. Water should be added when the hole is half full to help the dirt settle. Fill the hole to the top and carefully press the earth around the plant.

Water and Mulch

Water the new perennial plant. Soak the soil completely to ensure that the root ball is covered in water. Mulch should be applied 2 to 3 inches deep. Organic mulch, such as triple-ground-shredded hardwood bark or compost, is ideal for perennial plants. Over time, this kind of mulch decomposes, supplying nutrients and aiding in soil development. Do not cover the crown of perennial plants or pile mulch against their stems (this leads to rot). Perennial stems should have the mulch scraped back from them so that the plant in the hole creates a doughnut.

ADD IN ANNUALS

Annual plants

After perennials, the next best plants to have are annuals. You might have several young fruit and nut trees, as well as transplants of other plants like rhubarb and asparagus, but as a whole, your garden right now probably just looks like a bunch of twigs.

You should unquestionably add lots of annuals to your garden to fill up all the empty space while your perennials are still establishing themselves in the first few years.

Let's do a walkthrough on how to add annuals to your planted catalog.

Seeds or Plants?

You've set out your garden, but the important question still needs to be answered. Is it preferable to start your annuals from seed, or should you get the tiny six-packs of mature plants from the garden center?

There is no one right answer. It really depends on what works for your schedule and what you anticipate for the spring.

In general, it is more economical to start from scratch. A packet of seeds that yields enough blossoms for several flats can cost $2 or $3. Furthermore, since most greenhouses only carry the most well-liked types in six-packs, there is frequently more variation to be acquired from seeds than from a container of flowers.

However, many individuals would rather just purchase a flat of seedlings that have already sprouted and started to bloom in order to have the initial portion of the labor done for them. This is undoubtedly the simpler option because you save the hassle of having to figure out when and how to plant annuals from seeds.

Picking seedlings could be a good idea if you're planting annuals later in the growing season. The seeds' germination and growth will take some time. The season may already be almost over when they start to bloom. If you are going to start from seed, early in the growing season is the best time to do it.

Starting From Seed

Don't panic; starting from scratch can initially seem like an intimidating procedure. It's much simpler than it seems to plant annuals.

As a general rule, annuals fall into one of three major types, each with specific planting requirements.

Hardy Annuals

These annuals are remarkably hardy. As soon as the ground is no longer frozen enough to work in, you can plant them in your outside garden. Generally, you should plant them in your garden at a depth twice as long as the seed's diameter.

Even in the preceding fall, you can plant certain hardy annuals. When doing this, be sure to sow the seeds farther apart than you would in the spring and cover the soil with a thin layer of mulch to prevent frost damage. The benefit of sowing these seeds in the fall is that they will blossom in the spring much earlier.

Half-Hardy Annuals

Although not quite as durable as the hardy annuals, these flowers may nevertheless withstand some cold. After the risk of a hard frost has passed and nighttime temperatures are typically above 25 F (-4 C), plant these in your outdoor garden.

Tender Annuals

These are a little more sensitive and undoubtedly less resistant to extreme cold. Only sow these seeds after all potential frost threat has passed. These delicate little seeds will likely suffer negative effects from a frost. To extend your growing season, it's a good idea to start these seeds indoors while it's still cold out. After the last frost, you can harden off these annuals and plant them in their permanent location.

Starting From Plants

It's practically impossible to go wrong with this procedure because it's so simple. You can do this with no problem at all, regardless of your level of gardening expertise.

Before you even start digging, it's a good idea to spread all of your plants out about the garden while they are still in their pots to better understand where you want them.

The plant's stems, leaves, and blossoms should remain above ground when you dig a hole just big enough to cradle the plant's root system. Make sure to allow enough space between the holes you dig when planting many annuals.

Your young plant should be carefully removed from its pot or plastic wrapping. Never attempt to remove a plant by pulling on its stems. As an alternative, flip the packaging over while carefully catching the plant in your hand. Give it a little squeeze to get the dirt off the plastic package. The plant ought to just fall into your hands.

Put the plant into the hole you just made, and then carefully scoop loose dirt back around it to fill the hole and protect the roots. Your plant will stand securely upright if you pat the earth into place.

Always thoroughly water your garden after you're through planting—this aids in the plants taking root as soon as possible. Make sure the dirt gets a good soak. If you're going to mulch, do it right away. Even while it is not essential, it does assist the soil in retaining water and prevents weeds from sprouting, so I do recommend it.

The procedure is virtually exactly the same if you're asking how to plant annuals in a pot. Place the plant loosely in an empty pot while holding the plant's top lightly in your hand to maintain stability. Once the pot is full and the plant's roots are covered, add more dirt around it. Water it well to encourage the roots to expand in their new environment.

Basic Flower Care

The fact that once they are in the ground, annuals require little maintenance is one of their many advantages. They are typically excellent self-care practitioners. However, you can do a few easy things to maintain their health and look their best.

Watering

This is the simplest part. You don't even need to do anything if it rains frequently. But if it doesn't rain frequently, you'll need to intervene to provide your plants with water. Usually, just by glancing at a garden, you can tell if it needs water. Do the leaves appear to be wilting or drooping? Are the flowers dying and fading away? If the soil appears dry and pale and the leaves are not as bright green as they should be, it needs to be watered.

After examining your plant, you can also feel the dirt if you're unsure. It has to be watered if it is dry, hot, and crumbly. Healthy soil should be moist and dark.

But it's possible to overwater. Your plants should not be submerged under water. It should usually be adequate if you water your annuals once a day. However, if the soil still already feels damp, you don't need to water. If it's an especially hot day or a dry spell, check your garden several times per day for indicators that it needs additional water.

Weeding

Permaculturalists look at plants considered to be weeds differently than traditional gardeners. Rather than thinking, "I don't want this plant here, so it's a bad plant," we think, "While I don't want to use this plant in this spot, in a different situation, it could actually be beneficial." For example, dandelions are hated by many, but they are very healthy for the liver and packed full of vitamin C. Not only that, but their strong root system helps to aerate and break up compacted soil. So, while you may not want dandelions in the healthy beds you've developed, they will be very helpful in improving areas with poor soil.

When I say here to remove weeds, what I mean is to remove any plant growing in a spot where you don't want it to grow. This will keep your beds looking nice and make sure a plant you don't want can't steal nutrients away from the plant you do want.

You're in luck if you've added mulch to the perimeter of your flower beds. Most likely, you won't need to perform much weeding at all. Your annual plants are less likely to become choked by weeds thanks to the mulch.

Even so, take care to only remove the weeds and leave your flowers alone if you must weed. Make sure to remove the weed by the root when you are weeding; otherwise, it will quickly reappear. You can use your hands naturally or a specialized weeding instrument to do this.

Deadheading

This fancy term refers to removing the faded flowers from a plant. Your flowers will appear a lot nicer if you simply remove these with your hands. Deadheading also encourages fresh growth by forcing new flowers to emerge in place of the fading ones.

You'll soon be well on your way to having your own lovely annual garden that will endure all summer long if you use this planting advice for annuals!

There's no need to lament your garden after the fall weather turns colder and your annuals start to die off. The fall and winter months are the ideal time to start thinking about how you may make next season's garden even more colorful and attractive than it was this year.

COMPANION PLANTING

Next, we look at a nifty little trick we gardeners use to get a little more out of our plants; It's called Companion Planting. These companions are plant best friends with strengths that make up for each other's weaknesses. More specifically, it refers to the notion that some plants like growing next to one another very much. Others would rather not, however.

Fortunately, many more plants get along than those that don't, as you can see in the companion planting chart at the end of this article. Companion planting, as further defined by the University of Massachusetts Center for Agriculture, is the practice of growing two or more crops close to one another in the hopes that their proximity will aid in nutrient uptake, better pest control and lower pesticide use, improved pollination, and higher vegetable yields.

Beneficial Companions

A suggested companion planting combination typically offers some form of advantage to one (or both) of the plants or adds a bonus to your garden as a whole. I refer to these as "friends" in our companion planting chart. Some plants may have a close bond and rely on one another's inherent plant hormones to support growth or provide protection. Other partnerships could be as straightforward as one offering shade for another.

Incompatible Plants

On the other side, it is frequently advised against growing some plants close to others - their "foes," as I refer to them in our companion planting chart. Issues like stunted growth could result from those pairings.

Growing peas near onions or garlic is a common example of a bad companion planting combo. Why? Although this claim is frequently made, there isn't really much scientific data to back it up. Some hypotheses contend that peas are not "heavy feeders" while garlic and onions are. Their demands for fertilization are therefore incompatible. Onions and garlic don't like too much nitrogen in the soil, but peas do. Another theory is that the pea's delicate, shallow root system could be harmed by surrounding onion and garlic harvesting operations.

How Important or Scientific is Companion Planting?

Similar to the pea and onion example above, most of the "proof" for companion planting is based on theory or anecdotal evidence. This is particularly true in regards to the alleged incompatible flora. Let's just say that the scholarly writings I've tried to find that support companion planting with science fall short. However, it doesn't always imply that everything is false! All scientific hypotheses are developed by experimentation and first-hand observation. Companion planting is something to think about, in my opinion, since many gardeners have experienced its benefits in their own gardens.

Benefits Of Companion Planting

So how can you and your farm benefit from companion planting? Let me show you just how!

Companion Plants Attract Pollinators & Beneficial Insects

To draw pollinators like bees and butterflies to your garden, grow fruit or vegetable plants beside flowers like calendula or borage. Self-fertile veggies include tomatoes, green beans, peppers, and peas. They can therefore produce fruit without the assistance of a pollinator. In contrast, many other plants such as berries, fruit trees, melons, okra, cucumber, and squash all depend on pollinators to grow.

Whether there is a fruit or not, I always recommend planting for pollinators! All the assistance you can give is helpful.

But we don't just want to support good guys in our gardens, like bees and butterflies. Other small beneficial insects like lacewings, parasitic wasps, hover flies, predatory mites, or ladybugs can be attracted by companion planting with herbs like basil, fennel, cilantro, sage, and dill by providing a shelter, food source, or other attraction. They all contribute significantly to organic pest management.

Companion Planting for Pest Control

▷ Pest Deterrents

Some companion plants prevent pests as a result of their distinct nature. Some plants have traits or compounds that are unfavorable or harmful to certain garden pests.

As an illustration, hot pepper plants have a chemical in their roots that are good at fighting off and avoiding root rot ailments like fusarium rot. As a result, neighboring plants in the same planting bed are protected from developing root rot. Basil and other fragrant herbs like dill, chives, and cilantro also work as aphid deterrents. According to reports, nuisance insects are kept at bay by the strong scent of marigold blooms.

▷ Predatory Insects

As we've already seen, certain companion plants draw helpful insects, some of which are predatory by nature. Our little ladybug and lacewing friends devour aphids, mealybugs, and other soft-bodied nuisance insects with ferocity. By placing their eggs and larvae on pest caterpillars, parasitic wasps help control the population of these noxious insects. Even though they don't specifically target the good insects, spiders and praying mantises also eat nuisance insects.

▷ Trap Crops

While those divert the pests away from your fruit and vegetable plants, other companion plants DO attract pest insects. These are referred to as sacrificial or "trap crops." A great illustration of a trap crop is nasturtium. While the adjacent plants are frequently spared, nasturtium draws both aphids and cabbage worms. Trap crop plants should be removed from the garden when severely affected. Nasturtiums are a favorite of bees and hummingbirds since they also offer pollen and nectar.

Companion Planting for General Garden Health

The flavor of life is variety! Variety is often preferred in a garden, in addition to the advantages of companion planting that we have already described. Each distinct plant can serve a purpose and be valuable. Some can be consumed. Others draw or keep away insects. Many are likely more medicinal than you realize! Furthermore, many companion plants are stunningly lovely.

Polyculture, or growing multiple crop varieties together in a small area, includes companion planting. Companion planting and polyculture boost your garden's biodiversity, which is a coveted accomplishment in organic gardening. A thriving biodiverse garden (a mini-ecosystem) has a lower risk of being overrun by disease or pests than typical agriculture does. There is less demand for pesticides and other chemicals as a result. Diverse gardens also have stronger immune systems that can withstand environmental stresses like drought, heat, or extreme cold.

Companion Planting Friends and Foes

Plant	Friend	Foe
Asparagus	Basil, Nasturtium, Parsley, Strawberry, Tomato	Onion, Garlic, Potatoe
Beans (Bush)	Beet, Celery, Corn, Nasturtium, Peas, Strawberry	Garlic, Onion, Leek, Chive, Fennel
Beans (Pole)	Carrot, Celery, Chamomile, Corn, Pea, Potato, Spinach	Beet, Broccoli, Brussel Sprout, Cabbage, Chive, Onion
Beet	Beans (Bush), Cabbage, Lettuce, Onion, Garlic	Beans (Pole), Dill
Broccoli, Cabbage, Kale	Chamomile, Dill, Garlic, Marigolds, Mint, Rosemary, Sage	Squash, Strawberry, Mustard
Corn	Amaranth, Cucumber, Parsley, Pea, Potato, Sunflower	Tomato, Cabbage
Dill	Cabbage, Cucumber, Corn, Lettuce, Onion	Carrot, Lavender, Tomato
Eggplant	Marigold, Peppers, Nasturtium, Spinach	None!
Garlic, Onion, Chives, Leeks	Beet, Cabbage, Carrot, Cucumber, Tomato, Thyme	Peas, Sage

Plant	Friend	Foe
Lettuce	Beet, Carrot, Chamomile, Radish, Rosemary, Thyme	Celery, Parsley, Onion
Melon	Corn, Marigold, Squash, Sunflower, Tomato	Potato
Mustard	Artichoke, Beet, Celery, Garlic, Rosemary	Potato
Oregano	Basil, Cabbage, Cucumber, Pepper	None!
Peas	Beans, Carrot, Corn, Spinach, Strawberry	Chives, Onion, Potato
Potato	Bean (Bush), Cabbage, Carrot, Marigold, Onion	Asparagus, Pumpkin, Sunflower, Squash, Turnip
Spinach	Celery, Cilantro, Onion, Peas, Strawberry	Potato
Strawberry	Garlic, Onion, Spinach, Thyme	Cabbage, Broccoli, Kale
Sweet Potato	Bean (Bush), Beet, Dill, Parsnip, Potato	Squash
Thyme	Eggplant, Potato, Strawberry, Tomato	None!
Tomato	Asparagus, Basil, Carrot, Garlic, Marigold, Mint, Nasturtium Onion, Peppers	Bean, Cabbage, Corn, Dill, Pea, Potato

Tips to Get Started with Companion Planting

1. Flowers and herbs make for some of the greatest and simplest companion plantings for your veggie

garden. You can use calendula, nasturtiums, basil, cilantro, oregano, parsley, thyme, and rosemary everywhere, as none of these are known to be "foes."

2. Remember to use excellent plant spacing, including for companion plants, when you are growing a variety of things in one area because overcrowding plants can negate the advantages of companion planting that you were hoping to achieve in the first place! Overcrowded plants compete with one another for nutrients, water, sunlight, and airflow. Diseases like mildew and blight are more likely to affect them. It is also simpler for diseases and pests to spread between them. The good news is that, for the most part, companion plants and herbs respond well to pruning, so feel free to do so whenever necessary to make room for your vegetables.

3. Create a plan. Make a plot plan for where you want to plant the seedlings before going outside and throwing them in a bed carelessly. You can then use the companion planting chart as you go. You won't be as likely to run out of space and have to place two enemies close to one another. It doesn't even have to be finalized, but having a strategy definitely makes me more organized.

Companion Garden Layout

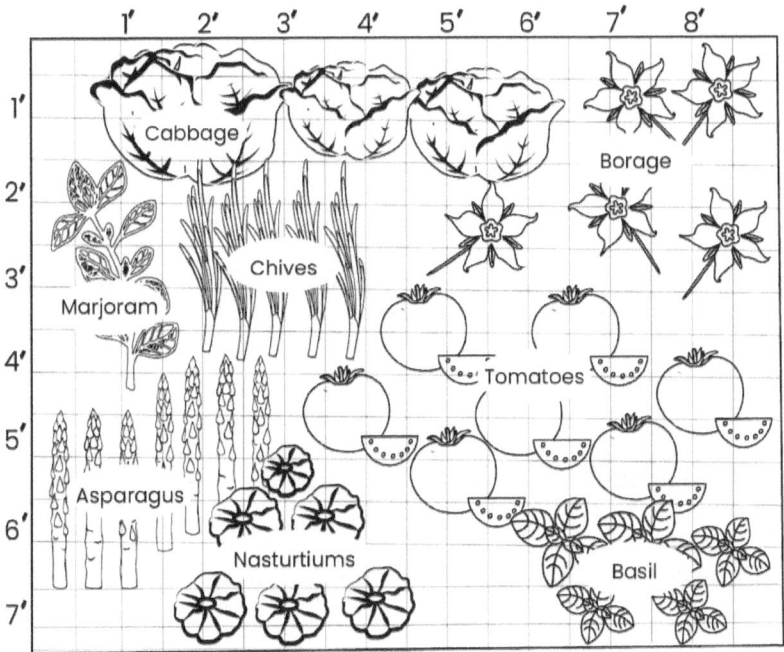

PLANT GUILDS

A permaculture guild is similar to companion plants, but it takes it a step further. It's a group of benevolent plants that decreases the effort required of the gardener while simultaneously benefiting the environment, wildlife, and other living things around it. A healthy and low-maintenance guild benefits from the guild's provision of disease prevention, fertilization, and pollination.

Guilds for the Home Garden

Many plants work together in guilds to achieve a stable coexistence where the garden is mulched, the soil is nourished, the pests are controlled, the pollinators are attracted, the nutrients are gathered, and the cultivators are fed.

Let's take a look at an example of a guild. We'll start with a traditional combination of companion plants like carrots and onions. Both carrots and onions enjoy growing next to lettuces, which might add a wonderful ground cover to the mix, keep the soil moist and the soil life safe, and produce a steady harvest in the meantime. Peas pair well with carrots but not so well with members of the onion family; therefore, vining peas could be trellised next to carrots but apart from onions to aid in nitrogen fixation and to provide a vertical component to the arrangement. An additional taller element in the arrangement, possibly the centerpiece, could be rosemary, a fantastic perennial herb that helps ward against pests. A pest-repelling upper layer, root crops, edible ground coverings, and nitrogen-fixing vines to shade the lettuce are now present. We have greatly increased diversity and functionality.

Another example is corn, beans, and squash--commonly referred to as the three sisters. While these three operate well together and might be regarded as a guild as is, we might be able to improve it. Beans are able to use the corn's stalk as a trellis for them to grow on while also providing nitrogen to nourish the other plants. The large leaves of the clambering squash plants form a ground cover that retains moisture. Comfrey may be an excellent addition to this mixture, providing a chop-and-drop mulch, a deep-rooting nitrogen accumulator, and a pollinator attractant. Sunflowers may be effective insect deterrents and providers of nutrient-rich food, but beans don't respond well to their allelopathic properties (so be aware). If you don't have enough room to plant them away from the beans, amaranth will work better.

The next step is to try to connect the dots logically and experiment. Knowing what to think about when creating a guild can help you achieve that.

Important Considerations for Home Garden Guilds

Although companion charts are excellent resources, they are only one of many elements that go into creating a successful guild. Understanding why particular plants pair well with one another and what factors to take into account when attempting to build new groupings is crucial. This is where having a foundational understanding of guild building helps. The next step is to think of and try out pleasing combinations. Consider replicating a successful assemblage and thinking about subtle ways it could be expanded or improved. But we have to get started somewhere.

What occupies the center?

There is often the main crop in guilds. The solution is evident when constructing around apple trees, but when constructing around vegetables of fairly equivalent size, the yields and value levels become comparable. Choosing the veggie that is either the largest or the most appealing may be helpful. Tomato plants don't always offer much to the nearby plants, but they offer a lot to us. They have a ton of kitchen versatility. They are the foundation of numerous sauces and soups and are essential for a vegetarian sandwich. They arrive in great numbers, and we want them for that reason. Let's choose something that most gardeners enjoy for this exercise: the tomato.

What is that smell?

Strong-smelling plants frequently repel pests while attracting pollinators. There are many different types of culinary herbs, and most of them get along well with other plants. One of these is basil. It not only gives off a smell that keeps pests away from tomatoes but also goes well with tomatoes on a plate, making it ideal for cooperative harvesting. Fresh basil adds a lot of nutritional and therapeutic benefits to food and a fantastic flavor. Additionally, it develops into a good-sized bush that takes up the area left by the high tomato vines.

What shields the soil?

Mulching is essential for successful gardening because it contributes organic matter, reduces erosion, maintains soil life, and keeps everything moist. While straw or dried grass clippings can be used initially, it's really nice to have a living mulch because it not only protects the ground, but also adds more plant variety. Nasturtium is a lovely, widely-spreading plant that covers the ground and produces both lovely edible blooms and tasty edible leaves. A poor man's caper can be made even from the seeds. Additionally, nasturtiums are supposed to enhance the flavor of tomatoes.

What causes fertility?

It's crucial to always take a system's soil replenishment into account. Although tomatoes are relatively voracious plants, they don't really get along with nitrogen-fixing legumes. Therefore, a different plant will solve this problem for you. Usually, comfrey is the solution, but borage is a better option for this guild. It works wonders at keeping tomato worms away, is thought to enhance tomato flavor, draws pollinators (particularly bees), and replenishes the soil's minerals. Also edible and frequently compared to cucumbers are the leaves and blossoms. It is an annual that self-seeds and can be planted directly in the ground. Borage is responsible for fertility, among other things.

What is at the root?

The plants that complement one other in cuisine frequently work well together, which is something I frequently discover when companion planting and guild building, for example:

- carrots with peas
- basil with tomato
- chives and apples

Okay, so it doesn't always work, but I enjoy that I can and should grow things that I eat with other foods frequently. Garlic takes a long time to grow (about nine months), but the garlic chives are still quite flavorful and may be harvested much sooner. Tomatoes and garlic go together extremely well. Additionally, the garlic bulbs prevent blight by working underground. When planted properly, garlic can be a perennial crop that requires very little maintenance and doesn't take up much land.

MULCHING

We now look at a disposable, indisposable tool in the permaculture business: Mulch. It is a biodegradable organic substance that is spread on top of the soil. It resembles a forest floor with a covering of protecting leaf litter in the permaculture garden. A bare patch of the ground indicates damaged ground in the natural world.

Mulching can make one question whether the work is worthwhile. Since many gardeners are time-poor, they hardly have time to plant and harvest, let alone maintain everything weeded.

Mulch can improve the ecology and cut down on time needed for other tasks like weeding, watering, fertilizing, and insect management, even though it may appear like an additional item on the to-do list.

In addition to discussing the best materials to use and when to use them, let's look at some of the advantages of mulching.

The Advantages of Mulch

Your mulching efforts will pay off in a variety of ways. They include:

- holding onto moisture
- preventing erosion
- producing humus, the foundation of topsoil
- fertilizing
- eradicating weeds
- creating an appealing top dressing

How to Use Mulch Relative to Your Climate

Mulching has several advantages, but how you go about it greatly depends on your climate and the time of year. Everybody's experience with this exercise will be unique.

For instance, gardeners in hot, dry climates should cover their plants with a thick layer of mulch to retain moisture and shield plants from the sun's heat.

By using a thin layer of mulch, gardeners in cold, rainy areas can prevent the soil from being washed away during heavy rains while yet allowing excess moisture to drain. In fact, mulching too much in this climate can lead to fungus problems. Slugs and snails may be drawn to it as well.

As a result, it's crucial to mulch with purpose and adjust to the changing seasons.

Mulch Types and Their Uses

Let's go over the many kinds of mulch and how they might be used in a permaculture garden while we look for natural ways to simultaneously offer additional benefits.

In light of this, I prefer to concentrate on employing mulch made from plants and unpaid materials.

You might be shocked to learn that many of the materials frequently marketed as good for gardens can actually be rather harmful. Learn how to choose secure materials for mulching naturally by reading on.

▷ Living Mulch

Planting annuals or perennial living plants beneath the main crop can assist in controlling weeds, holding onto moisture, stopping soil erosion, and providing a habitat for important insects.

Due to their entire roots, they also provide a food source and habitat for beneficial soil organisms.

Look for annual plants to serve as live mulch in the food garden. In the meantime, use perennial plants as live mulch underneath edible perennial crops, such as fruit trees.

I wouldn't, for instance, grow the perennial herb comfrey in my food garden. This is because the soil in a vegetable garden is frequently disturbed during planting and upkeep, which is bad for perennial plants' roots.

A Few Examples of Living Mulches

Annual Mulches for the Vegetable Garden:

- Nasturtium
- Calendula
- Borage
- Sweet alyssum

Perennial Mulches for Perennial Crops

- White clover
- Rhubarb
- Oregano
- Comfrey

Will a living mulch suffocate the main crop?

In other words, if you use the wrong plant in the wrong location, a living mulch might suffocate a crop.

As we learned above, there are several climate and seasonal circumstances where mulch should be used differently. Therefore, it's crucial to adjust the living mulch in accordance with the current circumstances.

For instance, plant a compact living mulch to completely cover the soil while growing a living mulch in full sun and a hot, dry region where you would typically utilize a heavy layer of mulch.

Meanwhile, plant a living mulch in a cool, damp region (or in the shade) with some spacing between the plants to let excess moisture evaporate and allow each plant to spread out and get more sunlight.

Annual crops often have not acclimated to such conditions, yet many wildflowers have evolved to flourish when grown in competition with one another in a grassland.

When in doubt, keep the number of living mulch plants in the vegetable garden to a minimum and look for companion plants that won't shadow the problematic produce.

The living mulch's root system and any nearby vegetable crops should also be taken into account. Different plants have various root systems, some of which are shallow and fibrous and others which are narrow and deep. The two are often more compatible because they don't fight for nutrients when the nearby plant's roots are of a different type.

Swiss chard and sweet alyssum are two of my favorite veggie and living mulch pairings. I found that because these companion plants' roots are so different from one another (one is shallow and fibrous, the other is deep and taprooted), they don't compete with one another for nutrition. Instead, they support each other.

I also enjoy growing comfrey beneath fruit trees for a similar result. Fruit trees with shallow roots and comfrey plants get along well.

Try several combinations to determine which ones you like most and which ones suit your garden's conditions.

▷ Green Mulch

To smother weeds and provide crops a nutritional boost (fertilizer), green mulch is the use of live plant material.

This kind of mulch is often known as "chop-and-drop" mulch since the green plants are frequently planted exactly where the green mulch will be applied. Mulching on-site saves work by eliminating the need to drag wheelbarrows of organic material into the area.

Consequently, you might, for instance, regularly remove living mulches throughout the season and scatter them directly on the ground. You might keep them short to maintain their ability to anchor and safeguard the soil while keeping them from shadowing the primary crop. If you don't want your green mulches to go to seed, this measure can prevent them from doing so.

Typical Green Mulches

The plants listed below are a few more that make excellent green mulch in addition to the living mulches mentioned above.

- Yarrow
- Chickweed
- Chives
- Dandelion
- Parsley
- Plantain
- Purslane
- Rhubarb leaves

Did you realize that a lot of these plants are regarded as weeds? You might be wondering why I listed common weeds like plantain, dandelion, and chickweed as suitable mulches.

Because they grow quickly, anchor many roots in the ground to assist the soil's ecology in taking off, and produce a lot of green plant matter, weeds can aid in the healing and enrichment of damaged soil.

Uses for Green Mulch Plants

Leave the roots of green mulch plants intact so they can continue to provide food for healthy soil creatures. Cut the green material roughly into 3-inch pieces, then spread the green mulch over the soil. Make sure it doesn't get into contact with the nearby garden crops directly.

Green mulch can be topped with a layer of woodchip or leaf mulch for a more aesthetically pleasing appearance.

▷ **Leaf Mulch**

Did you know that, on average, trees' leaves contain twice as many minerals as manure, measured in weight? Fall leaves are an excellent mulch and soil improver in one!

In the garden, leaf mulch is a lovely top-dressing that also aids in moisture retention. You can pick up the leaves off the grass, chop them up with the lawnmower or leaf mulcher, and then store them in wire containers for the entire year.

Leaf mold, or composted leaves that have been sitting for two to three years, is a great soil conditioner.

▷ **Wood Chips**

Local tree services can provide wood chips frequently without charge (giving the delivery person a tip might result in additional deliveries in the future!). Use wood chips as a mulch around perennials, but make sure they never come into contact with the plant's stems or trunk; otherwise, the plant will become too moist and may start to rot.

A soil conditioner's gold mine is wood chips that have decomposed for two to three years. Use them generously in the vegetable garden or as a lovely top dressing under fruit trees.

Mulching is crucial for gardeners' long-term soil health, water conservation, and time efficiency. Plants thrive when the soil is healthy. Pest issues will decrease and soil, even if it starts out as poor quality, will gradually become more fertile.

MAXIMIZING SPACE

Many of us desire to cultivate plants for food and pollination even when we lack the luxury of ample acreage. There are various ways to make the most of your urban space so that you can produce food, herbs, and even trees. You can make your modest space flourish with imagination and commitment.

Think Vertical

Ordinarily, living in an urban area means having little open space. Go vertical is the answer! Keep in mind that fruits and vegetables such as cucumbers, tomatoes, squash, and beans are more than happy to grow vertically with suitable soil depth and staking. Also, take into consideration trailing and vining varietals. In some shade, leafy vegetables typically grow nicely.

Use trellises, window boxes, hanging baskets, and shelves to arrange plants in layers. Water hanging plants more regularly than other types of plants because they tend to dry up more rapidly. Urban permies are known to have pots on most of the stairs leading up to their apartment doors.

Containers

You may choose to make containers that best fit your space thanks to the versatility they offer in terms of size and shape. By pairing similar plants, you can make the most of your containers.

In order to have blossoms all year long, flowering plants can be planted in layers within pots, with summer bulbs at the bottom, perennials in the center, and annuals at the top. It is possible to cultivate many fruits and some fruit trees in containers. A 66-square-foot edible rooftop garden maintained by Marie Viljoen was maintained nearly entirely in containers.

Rooftops

A lot of sunshine and surface area are available on rooftops. Try any combination of vertical planting, arbors, trellises, and containers to create your garden if you have access to a rooftop.

Rooftops can also be a terrific place for bees to live, but you should be careful of tar-covered roofs because they emit more heat than normal and may be too much for your bees.

Water

Consider purchasing or building a rain barrel if you have the space—which you probably have more of than you realize. Your plants will require a lot of water, and plants in pots are especially susceptible to drying out in hot weather. Make sure your rain barrel is protected from mosquitoes and other water-loving insects with a screen or other another protective device.

Compost

An efficient composting system doesn't require a lot of space to operate. In addition to producing odors that are unappealing to you and your neighbors in confined quarters, the conventional cold compost bin system requires time to decompose.

Constructing tumblers from recycled materials is simple, and regular spinning promotes aeration and expedites composting. Worm bins are also an excellent choice for interior composting and, as long as you don't introduce any odorous plants (such as onions or garlic), won't produce any smells.

Beekeeping

This is not just for large farms or homesteads. With back-yard beekeeping, many urban homesteaders have found success. Bees can thrive in the backyard of a row house or townhouse and can be very content on tar-free rooftops or balconies. Be sure to verify your local rules before constructing your hive, as beekeeping is not often permitted in major metropolitan areas.

Ensure that you give a water source (perhaps in the shape of a water garden) to prevent bees from seeking another water source, such as in a neighbor's yard.

Fowl

A small flock of successful poultry can be kept in a city, although they do need a little more room than some other garden endeavors. Again, before you begin, ensure you comply with all applicable local laws. It's common that even if keeping chickens is acceptable, keeping roosters would be against the law due to noise regulations. To kill any potentially dangerous germs, chicken manure needs to cure for a few months if you intend to use your flock to contribute to compost.

Neighbors

Living in an urban area means that our neighbors are always nearby. Being respectful is vital because some of your back-yard objectives might not coincide with those of your neighbors. Be sure to talk with them before adding your backyard chickens or bees, and never forget that a gift of produce or a jar of honey can go a long way toward keeping everyone happy.

I put a lot of time and energy into writing this book because I really want to show people who are dreaming of a permaculture garden that it doesn't just have to be a dream. You can do it. One thing I really like about this book is its order of progression. You'll find that every step, each chapter, builds upon the last. I was about to express how this is the most important chapter because here, you learn about what to plant and how to plant it. However, the next chapter is arguably more important, if not equal.

Next, we look at the maintenance of the entire garden. You have done much work if you practiced with this book, but now you need the knowledge to create longevity. Proper maintenance is crucial to any farm, and I'm going to show you just how you can keep up yours.

8

STEP 7: MAINTAINING YOUR GARDEN

This may be the final chapter of your permaculture guide, but it's only the beginning of your new life as a permaculturist. You've learned about your space; you know what to plant and which animals to raise; you understand what it takes to grow crops; now, you must learn how to maintain it all.

This chapter walks you through the principles of garden maintenance. You'll learn to keep your farm healthy and productive so you can have a long and enjoyable permaculture journey. Let's get right into it.

COMPOSTING

Getting Started

First and foremost, you need to know the makeup of compost. Carbon and nitrogen are the two main components.

Nitrogen is derived from living things and should be utilized in smaller amounts to hasten decomposition. Table scraps, tea leaves, kelp, fresh grass cuttings, fresh leaves from the garden, flowers, cuttings from fruits and vegetables, coffee grounds, chicken manure, and seaweed are examples of this. While you are making compost, they contribute moisture. Your compost may become smelly, bulky, and slow to decompose if you utilize too many of these substances.

Carbon is found in brown matter, which promotes aeration and feeds the compost's microorganisms. This speeds up decomposition. The brown material consists of paper scraps, straw or hay, shrub trimmings, wood ash, outdated newspapers, dry leaves, corn cobs, dryer lint, stalks, cardboard, and pine needles.

Like an artificial fertilizer, compost is organic debris that has decomposed and now contains nutrients for plant growth. Unlike artificial fertilizer, compost also serves as a food source for soil fauna, including earthworms and microorganisms (bacteria and fungi), enhancing soil structure and fertility.

Improved soil structure allows roots to bury themselves more effectively, which enhances soil aeration, drainage, and infiltration. Compost offers a substrate to improve water and nutrient retention in sandy soils. It also helps lessen the effects of low organic matter and fertility, compaction, erosion, and soil degradation in degraded soils.

This mix generally aids in improving the amount and quality of plant yields while rejuvenating and safeguarding soils.

There are three types of composting; Cold composting, Hot Composting and Vermicomposting. Let's take a closer look at what they all mean.

Brown Material (Carbon)	Green Material (Nitrogen)
Dried Leaves	Fresh Grass Clippings
Pine Needles	Vegetable Scraps
Twigs	Coffee Grounds
Cardboard	Fresh Leaves
Newspaper	Weeds (not gone to seed)
Straw and Hay	Fruit Scraps
Shredded Kraft Paper	Tea Leaves
Dried Shrub Trimmings	Kelp
Sawdust (untreated wood)	Flowers
Corn Stalks	Chicken Manure

Do Not Put In Your Garden Beds or Compost	
Dog, Cat, or Human Waste	Dairy
Weeds (have gone to seed)	Glossy Paper
Plastic or Styrofoam	Diseased Plants
Oils	Treated or Painted Wood

DIFFERENT TYPES OF COMPOSTING

In its most basic form, composting can be accomplished by building up a pile of wet food and garden waste and letting worms and microorganisms perform the work of decomposing it.

Cold Composting

This is the simplest method of composting and is usually done on a small scale. Let's say you have some kitchen scraps left over from making dinner. Just throw those scraps in a container, add some dried leaves as your brown material, and you have started to cold compost! Each time you have leftover scraps, you can add them to your container. Over time, you will gradually end up with more and more compost.

It is simple to cold compost because you don't have to turn it. Because the temperature is lower, there is less off-gassing of nutrients like nitrogen and carbon dioxide, and the soil biota may produce humus for a longer period of time throughout the extended maturation period.

Cold composting is very low maintenance compared to the other types; however, it often takes six months to a year or longer.

How to Make

1. Consider Compost Location

Composting is fairly simple and can be done in a shaded area of your garden or backyard if you have the space.

On the ground or in a bin, you can cold compost. Make a mound on the ground if you have the room or just don't want to spend money on a composting container. A bin is an additional choice if you have less room or wish to keep your compost contained. A circle of chicken wire or woven wire fencing that has been joined to itself at the desired circumference can also be used to create a simple container for your compost.

Cold Compost

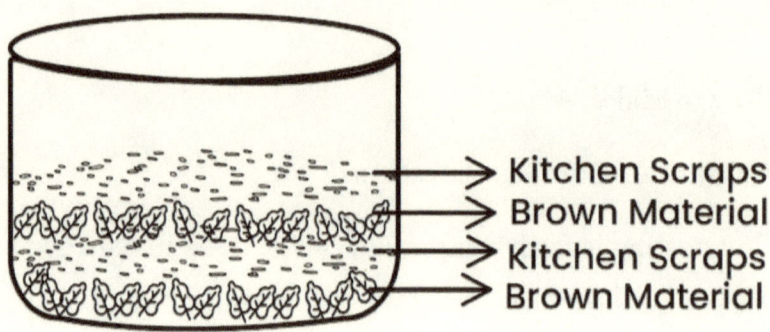

→ Kitchen Scraps
→ Brown Material
→ Kitchen Scraps
→ Brown Material

2. Prepare Your Composting Space

You can begin your cold compost once you've determined where it will be located. If you're doing this on a small scale, you can simply put one part kitchen scraps along with two parts brown material into a container anytime you have kitchen scraps. If you want a larger compost pile in your backyard, start with a six-inch layer of brown material, such as leaves, tiny branches, dried lawn clippings, newspaper, or torn cardboard placed on bare ground.

3. Fill In With Your Nitrogen-Rich Green Material

Place the second layer of brown material on top of the kitchen waste, covering it to a depth of 6 inches. You could add a second layer here (and cover with more brown materi-al), depending on how much green material you have, or you could stop at one layer. Maybe you don't have more kitchen scraps right now, but after a couple of weeks, you have accumulated more. Add them whenever you have them. Cold composting will work around your schedule. Always finish with the brown stuff on top.

4. Wait

Once you've covered your compost heap with a layer of carbon-rich brown material and aren't in a rush because you're cold composting, you can essentially just let it do its thing.

Continue adding layers of brown to green in the same ratio as before, or around two-thirds brown to one-third green. Add the green stuff in once or twice a week, covering it each time with brown.

5. Pick Up Your Compost

Your compost pile should be the same size as when you started, even though you have been adding to it, after 4-6 months (the length of time varies on rainfall and air temperature). Toward the end of the procedure, it may be 70–80% smaller depending on how quickly the decomposition process proceeds. This indicates that the materials have broken down effectively.

It's time to start enjoying the benefits of your compost. The amount of compost you can extract from your pile will depend on how quickly it decomposes, which in a cold compost depends greatly on local moisture and air temperature conditions. At the end of six months, you should have at least 4-5 gallons of compost if you are adding a gallon of material to your pile each week.

Compost should resemble a crumbly, dark brown substance that smells excellent and is moist. Nothing from what you composted that can be identified should remain.

Application

Compost can be added to the soil when repotting indoor plants, or it can be blended 50/50 with potting soil for beginning seeds. Before spring planting or right after harvest in the fall is the ideal time to put compost directly onto the soil of a garden bed or container.

Alternatively, you can put it straight into the beds as you plant annual or perennial flowers or bulbs, or you can mix it into the soil when you plant trees or shrubs. Compost can be used as mulch to prevent weed growth while also enriching the soil as it decomposes (particularly if it hasn't completely broken down yet). Compost can even be applied to your grass in the spring or the fall.

Advantages

The best thing about cold composting is that you can "set it and forget it." Consider the scenario when you don't care how long it takes to produce finished compost. Only in the spring do you intend to harvest the compost. Additionally, if the scent is not an issue because you have a lot of land, that's fine. Simply place all of your compostable kitchen waste on top of a mound and leave before a wasp sting or the stench knocks you out!

Hot composting

When you combine various organic materials (food scraps, grass clippings, manure, straw, weeds, etc.), mix them into a sizable pile, add some water, and voila! You have created a compost heap. Hot Composting maximizes microbial activity to produce compost in a shorter amount of time. The compost pile's warm temperatures encourage this action.

How to Make

1. Select a location for your heap

The ideal location is at the center of the garden, not in a remote area that is likely to be overlooked. Composting is hip; take pride in it and flaunt it. Compost piles can be significantly funked up to improve their aesthetic appeal. You can either leave this heap unattended or erect a structure out of wooden pallets, tin, timber, straw bales, or plastic units.

2. Gather the ingredients

You'll need sufficient supplies to create a pile that is three feet (one meter) high, three feet wide, and three feet deep. Two categories of ingredients are used:

- Materials rich in carbon (also known as browns), such as straw, hay, dried grass clippings, and shredded paper or cardboard
- Materials high in nitrogen (also known as greens), such as leftover food, fresh grass clippings, weeds, and animal manures

Hot Compost

3. Construct the heap

Alternate layers of greens and browns while adding a little water in between each layer. So there could be a layer of straw 4 inches (10 cm) thick, followed by a layer of food waste 2 inches (5 cm) thick, a layer of shredded newspaper, a layer of grass clippings, and so forth. Water is essential; each layer should be moist like a wrung-out dishcloth.

4. Cover the pile

When the heap has been constructed to the magic three feet size, cover it with some old carpet, black plastic, or straw and set it aside to cook. Check to see if it's hot a few days later. If nothing seems to be happening, the pile is probably too wet or too dry. Both are easy fixes. If it's too wet, spread the pile out, and the extra sun and wind it's exposed to will take away some of the moisture. If it's too dry, all you have to do is add more water. If that is not the problem, the pile is lacking in nitrogen (green material). Add some fresh grass clippings, kitchen scraps, or animal manure.

5. Rotate the pile

It all comes down to you, your energy, and your time. It is labor-intensive, but converting all those individual pieces into beautiful compost is feasible in less than 20 days. It is advised to turn it in at least once every couple of weeks. This effectively stirs the mixture and prompts the heap to heat up once more, allowing you to evaluate the compost heap's performance. You may make compost more quickly by turning it more frequently.

6. Beautiful Compost

When none of the original inputs are recognizable, the system is ready. You will have a rich, dark chocolate brown color with a soil-like texture. Hooray! All your time and effort paid off. Now you can put it to use in your garden.

Advantages

The speed of hot composting is its main benefit. Keep an eye on the "green vs. brown" ratios and turn it frequently. Results should be seen in a few weeks. Additionally, you do not need to stress too much about the items you add to the pile. A basket of acidic orange rinds won't make composting worms happy, but a warm compost pile won't care.

Vermicomposting and Vermicompost

Worms are referred to as "vermin" in Latin; hence, "vermicomposting" refers to composting with worms. Vermicomposting is the process of breaking down organic waste by employing worms, bacteria, and fungi. These creatures serve as nature's incredibly helpful decomposition agents for organic things. Therefore, vermicomposting is a method that speeds up nature's process of breaking down organic waste and yields a highly beneficial final product.

The final result of the vermicomposting process is called vermicompost, which is a mixture of organic wastes, decomposed worms, bedding materials, worm castings, and other decomposer organisms, cocoons, etc. Worm castings, one of the main ingredients of vermicompost, contain more nutrients and fewer pollutants than organic wastes do before vermicomposting. The abundance of water-soluble nutrients in vermicompost makes it a superior organic fertilizer.

This is my favorite kind of compost. Some might think it's my favorite because of how nutrient dense it is or because it doesn't smell. Maybe those reasons should be why it's my favorite, but really it's because I love watching those worms wiggle around!

How to Make

1. Using a three-compartment system is the best way to start. You can buy a vermicomposter online or make one yourself. If you have three 5-gallon plastic buckets and a drill, it's not too hard to make it yourself. Drill 1/8-inch holes in the bottom of two buckets. Stack all the buckets on top of each other with the hole-less bucket on the bottom.
2. In the middle bucket lay three layers of newspaper and lightly mist with water. Top that with about three inches (7.6 cm) of soil and plop in about 500 squirmy worms. Red Wigglers are the best kind for composting. On top of that, put about 2 inches (5 cm) of kitchen scraps. That's all for the middle bucket.

3. In the top bucket, add more worm food. This can be kitchen scraps, weeds pulled from your garden, spent flowers, or garden veggies that went bad before you were able to eat them. You don't have to add anything into the top bucket right away. It will take some time for the worms to eat through everything in the middle bucket, so don't worry if you don't have enough material to fill the entire vermicomposter when you set it up. You can add materials later when you have them. When your worm friends have eaten through their middle bucket food and turned it into rich compost, they will naturally start to look for more food and travel up through the holes to the top bucket and get to work turning that material into nutritious garden goodness. Once that happens, apply the contents of the middle bucket to your garden or place them in a different container to store until you're ready to use it.

4. At this point, your top bucket will move to be your middle bucket, and the now empty bucket on top is ready to be refilled with worm food.

5. Liquid will seep through these layers, and the bottom bucket is there to collect it. The technical term for this is leachate, but it is also called vermicompost tea or, my personal favorite, worm juice. This liquid is very beneficial to your garden and easy to use.

Simply pour this worm juice onto your garden beds, and your plants will thank you.

6. Lastly, you'll want to check on it to make sure it doesn't get too hot, cold, wet, or dry. It's widely accepted that the ideal temperature for vermicomposting is around 72 F (22 C), but between 55 F (13 C) and 80 F (27 C) is fine. If your worms are too hot, move them to the shade or add cool water. If they are too cold, add material around the outside of your vermicomposter to insulate them. For moisture, you don't want it to get dry and crumbly or wet and soupy in your vermicomposter. A slightly damp, moist condition is perfect. If it's too dry, mist water into the composter and if it's too wet, place it in a sunny or slightly windy location to dry out a bit.

Vermicompost

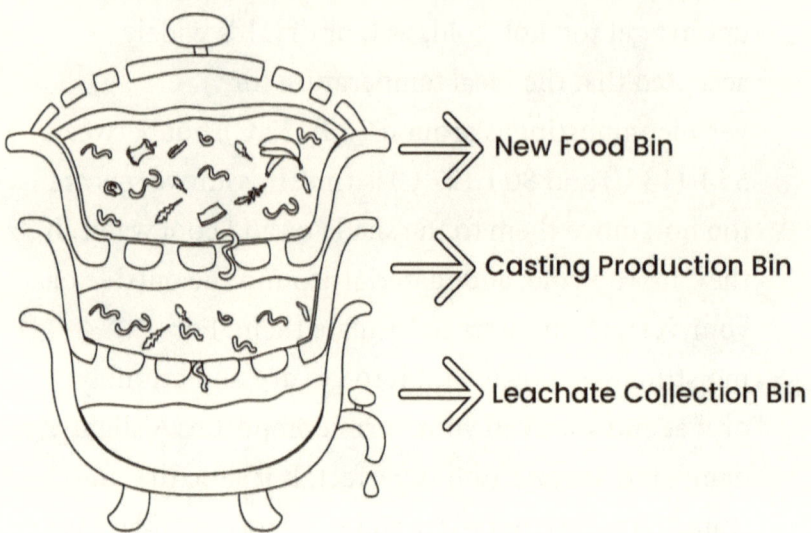

New Food Bin

Casting Production Bin

Leachate Collection Bin

Application

When the compost is prepared, it should be carefully applied to the soil to get the most out of it. The following methods for adding vermicompost to soil should be used:

1. Vermicompost can be applied like compost to the topsoil surrounding plants.
2. In general, vermicompost is twice as effective as other compost, so you only need to use half as much of it.
3. When applying the growing mixture to pots and containers, you should use roughly 25%

vermicompost with 75% soil rather than pure vermicompost.

4. Vermicomposting will naturally give you some compost tea, but you can make more worm juice if that is the way you prefer to use your vermicompost. Place the compost in water for a day and stir occasionally. If you want to take it up a notch, place an aquarium air pump in the bottom of the bucket for improved aeration. The next day, pour the contents of your bucket through a sieve to separate your vermicompost leftovers and worm juice. This liquid can then be poured or misted onto your garden beds.

5. In transplanting and bare-root transplanting situations, the compost tea can be sprayed on the plants.

6. The vermicompost needs to be prevented from drying out if you want optimum results. It can be kept in an air-tight container for a year or longer in storage.

Advantages

Vermicomposting is far superior to other forms of composting in four important ways.

1. Odor-free. If you take care of your worm bin properly, it ought to smell "earthy."

2. Inside or outside. Reside in a condominium? Dislike walking outside in the winter to the compost bin? When it's cold, do you want to continue composting at the same rate? Vermicomposting can take place in your yard, basement, garage, closet, or under your kitchen sink.

3. Speedy. A healthy, established worm colony can consume its own weight in food leftovers each day. This indicates that they have the ability to quickly transform garbage into riches.

4. Nutritious Results. Bins for composting produce more than just finished compost. The worms excrete a highly coveted material that gardeners refer to as "black gold." This dark substance, which resembles coffee grounds, contains both helpful microorganisms and natural fertilizer for the soil. These worm castings are one of the best soil supplements available.

STORING FOOD

Now that you've put in a lot of hard work to reap a bountiful harvest, the last thing you want is for your home-grown produce to spoil. Going back to one of the ethics of perma-culture from chapter 1, People Care, it's always a great idea to share from your abundance. Giving to family, friends, and neighbors is a great way to put your extra produce to good use while also making it easier on yourself. No need to prep for storage if you give it away! Giving back to nature is also always a good idea. If you're keeping animals, they may be thrilled to get your extras. Chickens love greens and think strawberries and blueberries are a special treat. Even if you intend to do something with food from your garden, but it starts to go bad before you get to it, put that in your compost pile. Nothing needs to go to waste; just find a different use for it.

While it's fun to give, the main reason you started your backyard food forest was probably to provide yourself and your family with delicious and nutrient-dense food. So, you're going to need to know how to store this food properly. I'll go over a few options.

Fresh Storage

I love vegetables that keep well in their natural state; they are the easiest. Onion and garlic are easy to box up and store in the basement. Each of these vegetables will remain fresh in storage for six to twelve months, depending on where and how you keep them.

Although there aren't many vegetables you can preserve fresh, garlic and onions are two excellent choices.

Once all of the bulbs have been removed from the soil, arrange them one by one on a raised surface (such as a sizable table or shelving rack) that receives filtered or indirect light. This might be in a garage with good ventilation, under a tree, or on a covered porch. Garlic will dry and be ready for storage in about two weeks. Onion takes about twice as long. You'll know they are ready when the outside is dry and flaky.

You won't need to clean off all that dirt for the time being because you'll clean them up when you use them in your food.

Refrigerator Storage

Another option for storing unprocessed vegetables is to use your refrigerator. One of my favorite things about growing vegetables in your backyard is that during the growing season you can just pop out and grab something to cook for that evening's dinner. But even after the first frost, you can still have easy access to vegetables that store well in the fridge with minimal prep work. Beets, carrots, spaghetti squash, and sweet potatoes are a few vegetables that keep well in the refrigerator so that you can have healthy and delicious veggies from your garden even in the middle of winter.

When you harvest these vegetables, wash off any dirt, and when they are dry, you can place them in plastic bags and store them at the bottom of your fridge until you're ready for them. Easy peasy.

Storage in a Freezer

For long-term storage, many vegetables can be quickly frozen. Some can be frozen uncooked, while others must first be blanched or steamed.

I advise purchasing a chest freezer if you plan to do a lot of freezing. The food quality lasts for around a year because it lacks the kitchen freezer's natural defrost cycle.

Kale, red peppers, and tomatoes are three of my preferred vegetables to freeze. All of those can be freshly sliced from the garden and placed right into bags or containers for freezing. You may simply take a handful and place it in the pan when you're ready to use them in a dish.

Fermentation

In the past few years, I've added the ability to ferment veggies to my arsenal of food preservation strategies. It's simple to duplicate the procedure with a variety of distinct vegetables once you get the swing of it and comprehend how it works.

Additionally, fermented foods retain a lot more nutrients than canned foods and provide healthy bacteria to your stomach. They can be kept in your refrigerator for up to a year.

Prepared Sauces

It's all about the sauce, goes the saying in our house. We frequently prepare one-bowl meals that include protein, a variety of roasted or sautéed seasonal veggies, and a foundation ingredient like rice, noodles, or quinoa. And (you got it!) a special sauce is added on top.

These sauces frequently contain herbs, which are particularly expensive to buy in the winter. Fortunately, they are really simple to cultivate, and it is simple to have more than you can eat.

That means that during the summer and fall, when the ingredients are plentiful, you may stockpile a freezer full of your favorite sauces. When it's time to make dinner in the winter, you can just defrost a jar and place it on top of a plate. Very simple.

The importance of proper maintenance of your garden cannot be understated. Mastering the art of composting, investing in efficient storage, and practicing simple gardening techniques could make all the difference. They may be small or big tasks, but their effect on the lifespan of your garden is invaluable.

I trust it all makes sense now. It has been a step-by-step guide all through this book, with Chapter 2 as Step 1. As you take it step by step, your understanding and fondness of the process grows. In time, an appreciation for the art slowly builds.

You may not recognize it, but completing the chapters of this book is already a win worth celebrating. If you choose to go ahead with your permaculture aspirations, your sustainable farming future looks promising.

CONCLUSION

Starting out in permaculture, I was greatly helped by a lot of resources. My aim in this book was to pay it forward and put all the most important information to get started in permaculture in one place so it can be easier for others. Is there more to permaculture than just what is in this book? Yes! So, so much more. But, do you need mountains of information to get started? No. Do you need to know everything about every single aspect of permaculture to have your own successful farm? No. My goal here was not to give a detailed description of the entire world of permaculture but to get you started. And if you follow the steps outlined in this book, you can definitely start and maintain a beautiful and delicious garden.

Starting off in chapter 1, we went over exactly what permaculture is for anyone out there who was unsure. Next, we dove right into the seven steps for how you can make your permaculture dreams into reality. We ventured into how you could familiarize yourself with your surroundings. You have to know what you have to work with before you can start doing the work. After that, I discussed how to choose the best plants and animals for your farm. There are a ton of options to start with here, so for anyone like me who suffers from option overload and can't make a decision when there are too many choices, I made a list of my favorites to start with. We then looked at how to design the layout of your site and ways you can provide your farm with much-needed water. Having all of that ready, building beds, and planting seeds should come as a breeze. Now you know the best ways and options to accomplish all that. And from there, it's just regular maintenance to keep up your lovely garden.

Permaculture is unique, and its methods are a bit unortho-dox, but it's not necessarily difficult. As with a lot of things in life, starting is the hardest part. But after you follow these steps and get your food forest set up, maintenance becomes the fun and easy part. And another great thing about it-- anyone can do it. Whether you have acres and acres of farm-land to work with, or you keep potted plants on the stairs going up to your apartment, anyone anywhere can do it. I know that with a bit of commitment, the challenges will pass, and with dedication, you will achieve a bountiful permaculture harvest!

I trust you've gained a ton from this little project of mine. If this book has helped you in your permaculture journey, **could you take a minute to leave a review on Amazon?** That will help me reach others who might want to gain the same things as you.

I hope you find the joy and fulfillment I and many others have found as you begin your permaculture pursuit. Happy planting!

Scan the QR code below for a quick review!